SOLAR CELLS
From Basics to Advanced Systems

McGraw-Hill Series in Electrical Engineering

Consulting Editor
Stephen W. Director, Carnegie-Mellon University

Networks and Systems
Communications and Information Theory
Control Theory
Electronics and Electronic Circuits
Power and Energy
Electromagnetics
Computer Engineering
Introductory and Survey
Radio, Television, Radar, and Antennas

Previous Consulting Editors

Ronald M. Bracewell, Colin Cherry, James F. Gibbons, Willis H. Harman, Hubert Heffner, Edward W. Herold, John G. Linvill, Simon Ramo, Ronald A. Rohrer, Anthony E. Siegman, Charles Susskind, Frederick E. Terman, John G. Truxal, Ernst Weber, and John R. Whinnery

Power and Energy

Consulting Editor
Stephen W. Director, Carnegie-Mellon University

Elgerd: *Electric Energy Systems Theory: An Introduction*
Fitzgerald, Kingsley, and Umans: *Electric Machinery*
Hu and White: *Solar Cells: From Basics to Advanced Systems*
Odum and Odum: *Energy Basis for Man and Nature*
Stevenson: *Elements of Power System Analysis*

SOLAR CELLS
From Basics to
Advanced Systems

Chenming Hu
Richard M. White
University of California, Berkeley

McGraw-Hill Book Company
New York St. Louis San Francisco Auckland Bogotá Hamburg
Johannesburg London Madrid Mexico Montreal New Delhi
Panama Paris São Paulo Singapore Sydney Tokyo Toronto

This book was set in Times Roman by Interactive Composition Corporation.
The editor was T. Michael Slaughter;
the production supervisor was Leroy A. Young.
The cover was designed by Mark Wieboldt.
Project supervision was done by Joan Stern of Interactive Composition Corporation.
Halliday Lithograph Corporation was printer and binder.

SOLAR CELLS
From Basics to Advanced Systems

Copyright © 1983 by McGraw-Hill, Inc. All rights reserved. Printed in the United States of America. Except as permitted under the United States Copyright Act of 1976, no part of this publication may be reproduced or distributed in any form or by any means, or stored in a data base or retrieval system, without the prior written permission of the publisher.

1 2 3 4 5 6 7 8 9 0 HALHAL 8 9 8 7 6 5 4 3

ISBN 0-07-030745-8

Library of Congress Cataloging in Publication Data

Hu, Chenming.
    Solar cells.

    (McGraw-Hill series in electrical engineering.
Power, and energy)
    Includes bibliographies and index.
    1. Solar cells.   I. White, Richard M., date
II. Title.   III. Series.
TK2960.H8   1983   621.31′244   83-975

CONTENTS

	Preface	ix
Part 1	**Solar Cell Fundamentals**	
1	**Overview**	**5**
	Box: How Much Oil?	6
1.1	How Solar Cell Systems Work	7
1.2	Types of Cells	8
1.3	How Much Power, and When?	10
	References	14
	Problems	15
2	**Nature and Availability of Solar Energy**	**16**
2.1	The Sun and the Sun-Earth Relative Motion	17
	Box: The Sun as a Fusion Reactor	17
2.2	Atmospheric Effects on Solar Radiation	20
2.3	Solar Radiation Measurement and Instrumentation	23
2.4	Geographical Distribution of Average Insolation	27
2.5	Effect of Collector Tilt	35
2.6	Summary	36
	References	36
	Problems	37
3	**Principles of Operation of Solar Cells**	**38**
3.1	Elements of Solar Cell Operation	39
3.2	Semiconductors	40

v

3.3	Light Absorption and Carrier Generation	42
3.4	Carrier Recombination	46
3.5	*pn* Junctions	48
3.6	Short-Circuit Current	54
	Box: Light Sensors: Photodiodes and Photoconductors	57
3.7	Efficiency	57
3.8	Factors Affecting the Conversion Efficiency	60
3.9	Summary	67
	References	67
	Problems	68

4 Materials and Processing — 69

4.1	Material Properties and Processing Techniques	70
	Box: Making an Integrated Circuit	71
4.2	Conventional Silicon Cell Processing	76
4.3	Processing Cadmium Sulfide Cells	85
4.4	Environmental and Other Considerations	86
4.5	Summary	89
	References	90
	Problems	90

Part 2 Concentrators and Complete Photovoltaic Systems

5 Concentration of Sunlight — 95

5.1	Solar Concentrators	96
	Box: f-Number of a Camera Lens	97
	Box: The Equation of Time	99
5.2	Economics of Concentrator Photovoltaic Systems	106
5.3	Concentrator Solar Cells	108
5.4	Cooling and Collection of Thermal Energy	110
5.5	Summary	111
	References	111
	Problems	112

6 Power Conditioning, Energy Storage, and Grid Connection — 113

6.1	Maximum-Power-Point Tracking	114
6.2	Principles of Maximum-Power-Point Trackers	116
6.3	Stand-Alone Inverters	118
6.4	Inverters Functioning with Power Grid	118
6.5	Costs of Power Conditioners	121
6.6	Issues of Energy Storage	122
6.7	Energy-Storage Technologies	124

6.8	Summary	126
	References	126
	Problems	127

7 Characteristics of Operating Cells and Systems — 128

7.1	Characteristics of Commercially Available Cells	129
7.2	Types of Applications	133
7.3	Operational Photovoltaic Systems and Devices	138
7.4	Summary	151
	References	153
	Problems	153

8 Economics of Photovoltaic Power — 154

8.1	Some General Rules	155
	Box: Ten Rules of Thumb	157
8.2	Cost Analyses for Photovoltaic Power	159
8.3	Utility Issues in Industrialized Countries	171
8.4	Issues in Developing Countries	173
8.5	Summary	174
	References	174
	Problems	174

Part 3 Solar Cell Improvements

9 Advanced Cell Processing Techniques — 179

9.1	Solidification and Thermal Activation	181
9.2	Forming Thin Self-Supporting Semiconductor Ribbons and Sheets	185
9.3	Forming a Semiconductor Layer on a Substrate	190
9.4	Use of Ion Implantation, Lasers, and Electron Beams	193
9.5	Optical Coatings, Contacts, and Encapsulants	196
9.6	Continuous Cell Production in an Automated Factory	197
9.7	Summary	200
	References	200
	Problems	201

10 Thin-Film and Unconventional Cell Materials — 202

10.1	Introduction	203
10.2	Amorphous Semiconductors	208
10.3	Miscellaneous Materials for Solar Cell Use	210
10.4	Prospects for Mass-Produced Thin-Film Cells	213
10.5	Summary	219
	References	221
	Problems	221

11 Variations in Cell Structures — 223

11.1	Review of Some Basic Structures	224
11.2	Unconventional Nonconcentrator Cells	225
11.3	Unconventional Concentrator Cells	227
11.4	Summary	235
	References	236
	Problems	236

12 Unconventional Cell Systems — 237

12.1	Multiple-Cell Systems—Spectrum Splitting and Cascade Cells	238
12.2	Thermophotovoltaic (TPV) System	244
12.3	Photoelectrolytic Cell	247
12.4	Satellite Power System	251
12.5	Summary	254
	References	255
	Problems	255

Appendixes — 257

1	Annotated Bibliography	257
2	Units and Relevant Numerical Quantities	259
3	Solar Spectrum—Air Mass 1.5	262
4	Abbreviations and Acronyms	264
5	Tabulation of Demonstrated Cell Efficiencies by Cell Type	268
6	Solar Cell Experiments	271
7	Computer Simulations of Photovoltaic Cells and Systems	276
8	Suppliers of Solar Cells	282
9	Overview of Some Operating Systems	285

Index — 291

PREFACE

This book began as a set of notes for two University Extension courses presented by the authors for heterogeneous groups that included college teachers and students, solar cell researchers, utility company engineers, government agency staff, a lawyer, investment counselors, and two people interested in installing a solar cell system in their Nevada ranch house. Success of the courses with such a diverse audience suggested making the material available, considerably reworked and augmented, in book form.

This book is intended both as a self-learning resource or professional reference and as a text for use in an undergraduate or first-year graduate course. A key feature of the book is modular organization. The twelve chapters, each containing problems and a reference list, may be read in almost any order, though they are grouped in three parts in order of increasing depth and complexity. The three parts discuss fundamentals, complete systems, and possibilities for improvements, respectively.

In Part One, Solar Cell Fundamentals, Chapter 1 gives an overview of photovoltaic power generation and its appeals. After the nature and the geographical distribution of sunlight is examined in Chapter 2, principles of operation of solar cells are developed in Chapter 3, starting from a discussion of the properties of semiconductors. For this the reader needs no more background than introductory college courses in mathematics and physics, although additional knowledge of semiconductors will be helpful. Chapter 4 concludes Part One with an examination of the materials and processing methods used in making conventional pn-junction cells.

Concentrators and Complete Photovoltaic Systems are the subject of Part Two. Concentrators (Chapter 5) and power conditioning and energy storage equipment (Chapter 6) may be used with solar cells for increased efficiency, convenience, and reduced cost. Examples of solar cell applications are given in Chapter 7, and the economics of photovoltaic electricity are discussed in Chapter 8.

Imaginative new approaches for further reducing the cost of solar cells are discussed in Part Three, Solar Cell Improvements. Chapter 9 describes new techniques for manufacturing semiconductors, and Chapter 10 examines thin-film approaches. Novel cell structures and unconventional cell systems are summarized in Chapters 11 and 12, respectively.

In addition, nine appendixes provide such information as an annotated bibliography, a list of solar cell suppliers, and a compilation of operational photovoltaic systems and their key design features. There is a table of recent laboratory cell efficiencies arranged by cell material and design. Several solar cell experiments that can be done with simple laboratory equipment are described in one appendix. Some readers may find useful the list of abbreviations and acronyms, and the discussion of computer simulations of solar cells and photovoltaic systems.

Perhaps no book can claim completeness in the coverage of a topic so interdisciplinary and fast-advancing; ours certainly does not. We have tried to stimulate thinking in addition to giving information, and hope to provide a foundation and framework upon which the reader can build further.

While writing this book, we have enjoyed the support and counsel of many friends whom we wish to acknowledge. Alex Kugushev of Lifetime Learning Publications suggested to us the idea of writing the book. Professor Ned Birdsall and many other colleagues urged us on. Joan Stern was the able editor. Bettye Fuller typed the manuscript. Finally, our families' support was invaluable.

Chenming Hu
Richard M. White

CHAPTER ONE
OVERVIEW

Components of a solar cell system, types of cells that exist or are proposed, and global power needs

CHAPTER TWO
NATURE AND AVAILABILITY OF SOLAR ENERGY

Characteristics of solar energy and ways of estimating, measuring, and collecting it

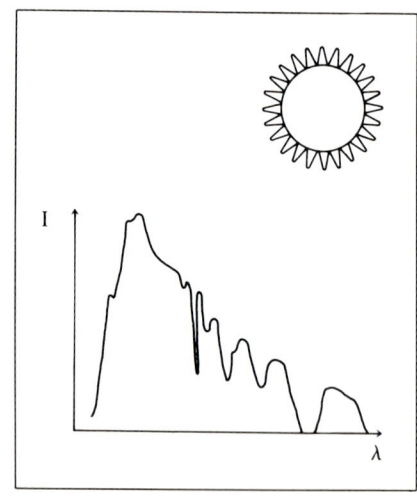

PART ONE

SOLAR CELL FUNDAMENTALS

CHAPTER THREE
PRINCIPLES OF SOLAR
CELL OPERATION
Nature of semiconductors and
the theory and efficiencies of
solar cells

CHAPTER FOUR
SOLAR CELL MATERIALS
AND PROCESSING
Making of conventional cells
with wafers cut from
single-crystal ingots

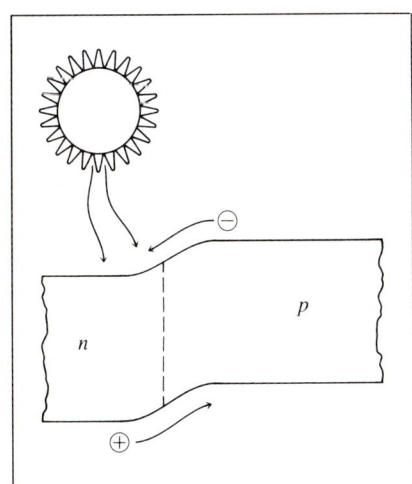

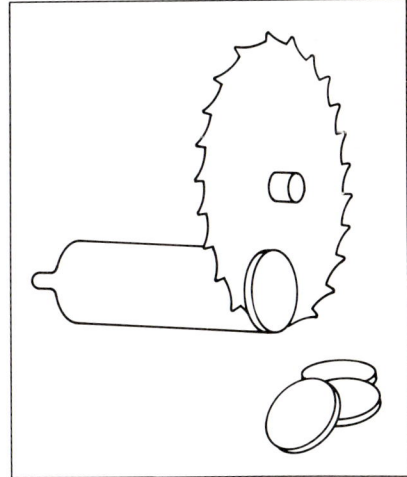

CHAPTER

ONE
OVERVIEW

CHAPTER OUTLINE

BOX: HOW MUCH OIL?
1.1 HOW SOLAR CELL SYSTEMS WORK
1.2 TYPES OF CELLS
1.3 HOW MUCH POWER, AND WHEN?
REFERENCES
PROBLEMS

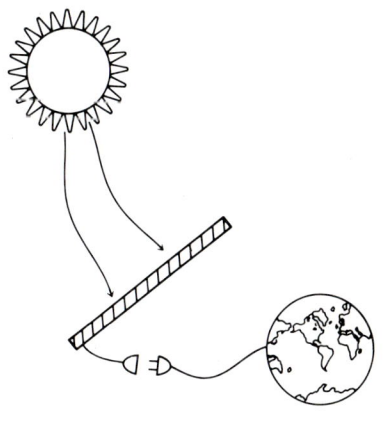

Solar cells are devices in which sunlight releases electric charges so they can move freely in a semiconductor and ultimately flow through an electric load, such as a light bulb or a motor. The phenomenon of producing voltages and currents in this way is known as the *photovoltaic* effect.

The fuel for solar cells—sunlight—is free and abundant. The intensity of sunlight at the surface of the earth is at most about one thousand watts per square meter. Thus the area occupied by the cells in a photovoltaic power system may be relatively large, and its cost must be considered in calculating the cost of the electricity produced. The primary factor that determines whether solar cells will be used to supply electricity in a given situation is the cost per unit output, relative to that of alternative power sources, of acquiring, installing, and operating the photovoltaic system.

Solar cells are already being used in terrestrial applications where they are economically competitive with alternative sources. Examples are powering communications equipment, pumps, and refrigerators located far from existing power lines. It is expected that the markets for solar cells will expand rapidly as the cost of power from conventional sources rises, and as the cost of solar cells falls because of technological improvements and the economies of large-scale manufacture. The first of these economic forces—the rising price of conventional sources, particularly those employing fossil fuels—continues automatically, in part because the resource is limited. The second—reducing the cost of electricity from solar cell systems—is the

---

HOW MUCH OIL?

It is sometimes argued that our energy problems would disappear if we just looked harder for oil. The inherent fallacy of this argument can be shown by the following example.

Suppose that geologists have been mistaken about our oil reserves, and have grossly underestimated the amount of oil that is in the earth. In fact, suppose that the earth consists *entirely* of oil, except for a thin solid shell on which we live.

The radius of the earth is 6.4×10^6 meters, and the volume of oil consumed worldwide in the year 1970 was 1.67×10^{10} barrels. You can show from these figures that if we continued to consume oil at that rate each year, the oil would last a comfortable 4.1×10^{11} years. But if consumption increased each year by 7 percent, as it did from 1890 to 1970, then the earthful of oil would be drained in only 344 years.

This example (Barnett, 1978) shows the devastating effect of an exponentially growing rate of consumption. For the record, the oil that is believed to be in the earth now will be used up in only a few decades if our present rate of consumption continues.

subject of worldwide research and development efforts today. To increase the economic attractiveness of the solar cell option, one or more of the following must be done:

- Increase cell efficiencies.
- Reduce cost of producing cells, modules, and associated equipment, and the cost of installing them.
- Devise new cell or system designs for lower total cost per unit power output.

1.1 HOW SOLAR CELL SYSTEMS WORK

The most important physical phenomena employed in all solar cells are illustrated schematically in Fig. 1.1. Sunlight enters the semiconductor and produces an electron and a hole—a negatively charged particle and a positively charged particle, both free to move. These particles diffuse through the semiconductor and ultimately encounter an energy barrier that permits charged particles of one sign to pass but reflects those of the other sign. Thus the positive charges are collected at the upper contact in Fig. 1.1, and the negative charges at the lower contact. The electric currents caused by this charge collection flow through metal wires to the electric load shown at the right side of Fig. 1.1.

The current from the cell may pass directly through the load, or it may be changed first by the power-conditioning equipment to alternating current at voltage and current levels different from those provided by the cell. Other sub-systems that may also be used include energy-storage devices such as batteries, and concentrating lenses or mirrors that focus the sunlight onto a

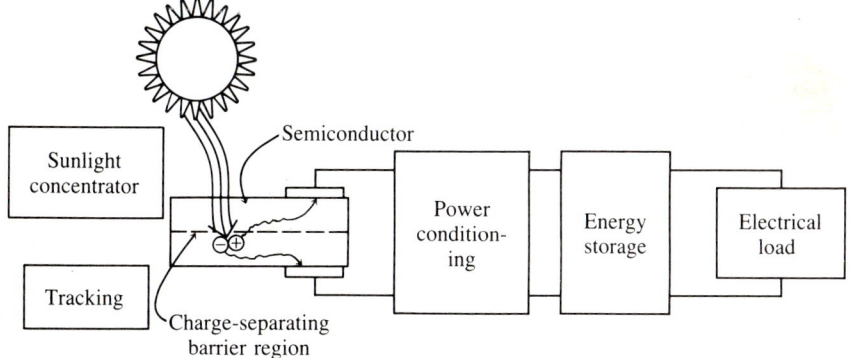

Figure 1.1 Sketch showing functional elements of solar cell system.

smaller and hence less costly semiconductor cell. If concentration is employed, a tracking subsystem may be required to keep the array pointed at the sun throughout the day.

1.2 TYPES OF CELLS

A glance at the recent literature (see "Annotated Bibliography," Appendix 1) shows that many different solar cell designs and materials are being studied. This situation is common in the early stages of a technical development, when many different approaches are explored. Incidentally, *terrestrial* use of cells to produce power is quite recent, even though the photovoltaic effect has been known since 1839. Silicon solar cells were first described in print in 1954, and solar cells have been used on most of the spacecraft launched since then. Cells for use in space are not discussed in this book because the hostile conditions those cells must withstand, together with the extreme reliability demanded, make the space cells far too costly and specialized for terrestrial use.

Although the details involve concepts of solid-state physics, chemistry, and materials science, some very simple observations underlie the different types of solar cells:

• Because silicon has been used so extensively for integrated circuits, its technology is well developed and it is a natural choice for use in solar cells now, while other approaches are being developed.

• Making thin-film or polycrystalline cells instead of single-crystal cells, which require extensive heating and careful crystal growth and slicing, may be economical in terms of both monetary cost and energy expended in the production process.

• Since focusing lenses and mirrors cost much less per unit area than do most semiconductors, it can be cost-effective to use "concentrator" systems in which sunlight is focused onto relatively small semiconductor cells.

• Since cells can be designed to work particularly well with light of one wavelength, it may be economical to split the spectrum and direct different portions onto cells optimized for those spectral components ("split spectrum" or "multicolor" cells).

• Since both available sunlight and the demand for energy fluctuate, cells providing inherent energy storage by electrolysis within the cell may be attractive ("photoelectrochemical" cells).

The cell types shown in Fig. 1.2 are arranged according to material and form of the semiconductor used and the degree of sunlight concentration employed. Some characteristics of the starting material or cell design are also indicated.

The cells produced in greatest quantity have been made of single-crystal

1.2 TYPES OF CELLS   9

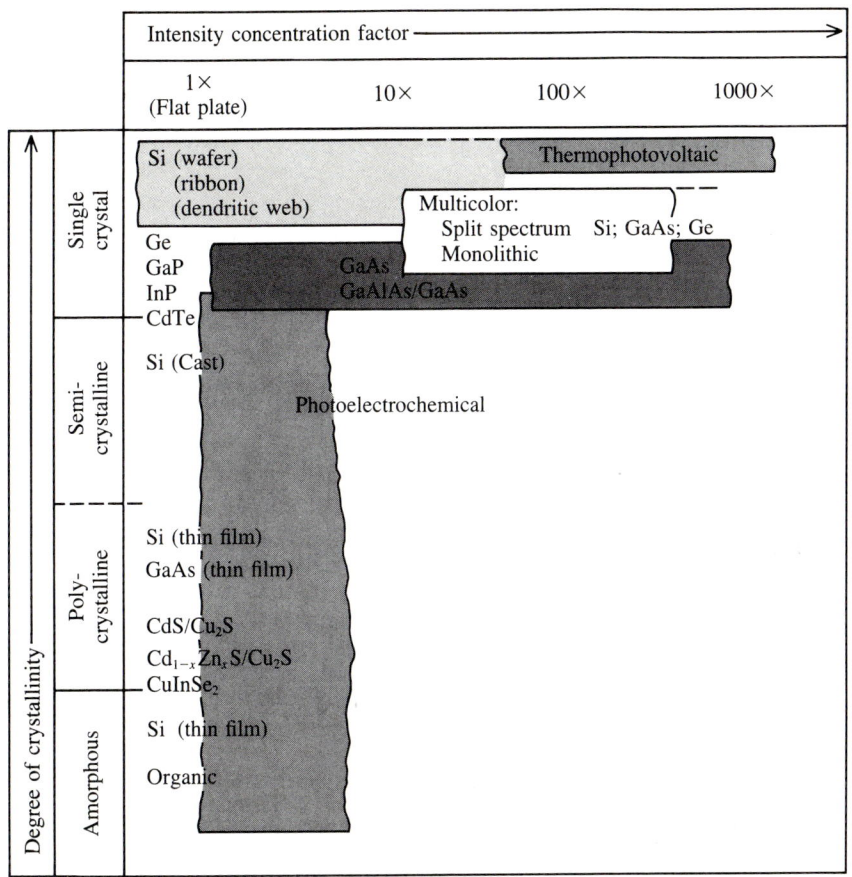

Figure 1.2 Types of solar cells, arranged according to degree of crystallinity of the semiconductor and degree of sunlight concentration used, if any. Single-crystal silicon and gallium arsenide, and thin-film CdS/Cu$_2$S cells are discussed in Part I of this book, while the other cell types and materials are discussed in Part III.

silicon, and used without sunlight concentration (upper left corner, Fig. 1.2). Silicon cells have been made of wafers sawed from large single-crystal ingots, and from thin ribbons or thin webs of silicon that do not require slicing. Of the many other single-crystal cell materials that have been studied, the compound semiconductor gallium arsenide has been most used in experiments because of its high efficiency and its ability to operate at high temperatures. Concentrator cell systems have been made with both silicon and gallium arsenide. Concentrator cell configurations differ from those for nonconcentrator use ("flat-plate" cells); concentrator cells must withstand higher temperatures and must have lower resistive losses because of their relatively higher cell currents. Since the split-spectrum cells involve several cells, one expects that to achieve low system cost such cells will be used initially with concentrating lenses or mirrors. Another concentrator cell involving spectral alteration in the interest of high efficiency is the "thermophotovoltaic" cell, which is illuminated by relatively long wavelength radiation from a plate heated by concentrated sunlight.

Most cells employ so-called *pn* junctions, that is, two adjacent regions of a semiconductor such as silicon that contain different impurities within them so they have different electrical characteristics. An alternative structure is the Schottky-barrier cell, in which a thin, fairly transparent metal film replaces one of the semiconductor regions of the *pn*-junction cell. Another promising cell design contains in addition a very thin insulating region between the metal and the semiconductor, forming the "metal-insulator-semiconductor" (MIS) or "metal-oxide-semiconductor" (MOS) structure.

Polycrystalline cells generally have lower production and material costs than do conventionally made single-crystal cells. Alternatives to the conventional methods of making single-crystal ingots have been developed, as will be discussed in Chap. 9. These include the edge-defined film-fed growth (EFG) and dendritic web cells. Thin-film cells, in which a semiconducting film is deposited on a substrate, include commercial cadmium sulfide cells, which actually have a *pn* junction between layers of cadmium sulfide and copper sulfide, and cells made of thin films of amorphous semiconductors. Experimental cells made from organic constituents are also under investigation. Studies predict that it will be possible to make thin-film cells that are efficient and inexpensive enough to become the cells of choice for many terrestrial applications.

1.3 HOW MUCH POWER, AND WHEN?

When one considers that the power rating of a modern central electricity-generating plant is typically 1000 megawatts, the present annual world production of solar cells of only a few megawatts seems very small indeed. (A *megawatt* is one million watts.) What reasons are there to think that solar cells may be important components in the world energy picture of the next few decades?

First, one must not judge significance solely on the basis of total power output. Even a 100-watt solar cell power supply in any one of the several million small villages in the world could be of enormous importance to the villagers, for whom it would provide power for water pumping, refrigeration, and communication with the surrounding world. Second, photovoltaic power generation is relatively free from the problems facing fossil-fueled or nuclear power plants—escalating fuel costs, disposal of wastes, disposal of heat, major concerns over safety, and potential modification of weather due to release of carbon dioxide. Third, photovoltaic systems are modular and can be installed near points of use and put on line quickly as the demand for electricity rises. These inherent advantages, plus the experience and expectations of a steady reduction in solar cell system cost, lead to the prediction that solar cell systems can make a significant contribution to the world energy supply.

Figure 1.3 summarizes energy supply and utilization in the United States in 1978. The unit standing for usage is the "quad," for quadrillion or 10^{15} Btu (British thermal units), a very large amount of energy. Several details are worth observing on this "spaghetti" diagram. In 1978, 54 percent of the energy was wasted, as can be seen by comparing the two energy amounts at the extreme right side of Fig. 1.3. Electricity production consumed 28 percent of the total energy input, and the production of electricity was on the average only about 33 percent efficient. The average rate of consumption of electrical energy in the United States in 1978 was about 260 GW (1 GW = 1 gigawatt = 10^9 watts, where 1 watt = 1 joule per second) and the total world rate of consumption of electrical energy was about three times the figure for the United States.

Predicting is always uncertain, but current U.S. government predictions say that when the cost of power from photovoltaic systems drops to $1.60 to $2.20/$W_{pk}$ for residential consumers, the annual U.S. market will be 3 to 10 GW_{pk}, and the electric utility market for photovoltaic power will range from 10 to 20 GW_{pk} per year. (The symbol $/$W_{pk}$ in this book means the cost in 1980 U.S. dollars of acquiring a solar cell array that produces 1 watt of peak electrical power when illuminated with sunlight at an intensity of 1 kilowatt per square meter. Because the sunlight intensity varies through the day, the peak output of a cell is, of course, greater than the power output averaged over a 24-hour day.)

When might photovoltaic electricity reach this price level? Active solar cell research and development is underway in industrial and governmental laboratories in nearly every developed country and in many developing countries. The detailed goals of the U.S. Department of Energy (DOE) effort appear in Table 1.1. The overall goal is to replace one quad per year of primary fuels by photovoltaics by the year 2000. In the recent past, commercial solar cell costs have dropped faster than the DOE goals, as shown in Fig. 1.4. Since 1980 the price of oil has actually fallen, and federal funding for photovoltaics has been cut back in the United States. Funding for energy research in several countries has increased (examples are West Germany, Japan, and Italy), and it is hoped that the industry will maintain the momentum for achieving the cost goals of Fig. 1.4. If the goals are met or exceeded, vast quantities of solar cells will be manufactured and used.

The anticipated rapid growth of solar cell use raises questions about the

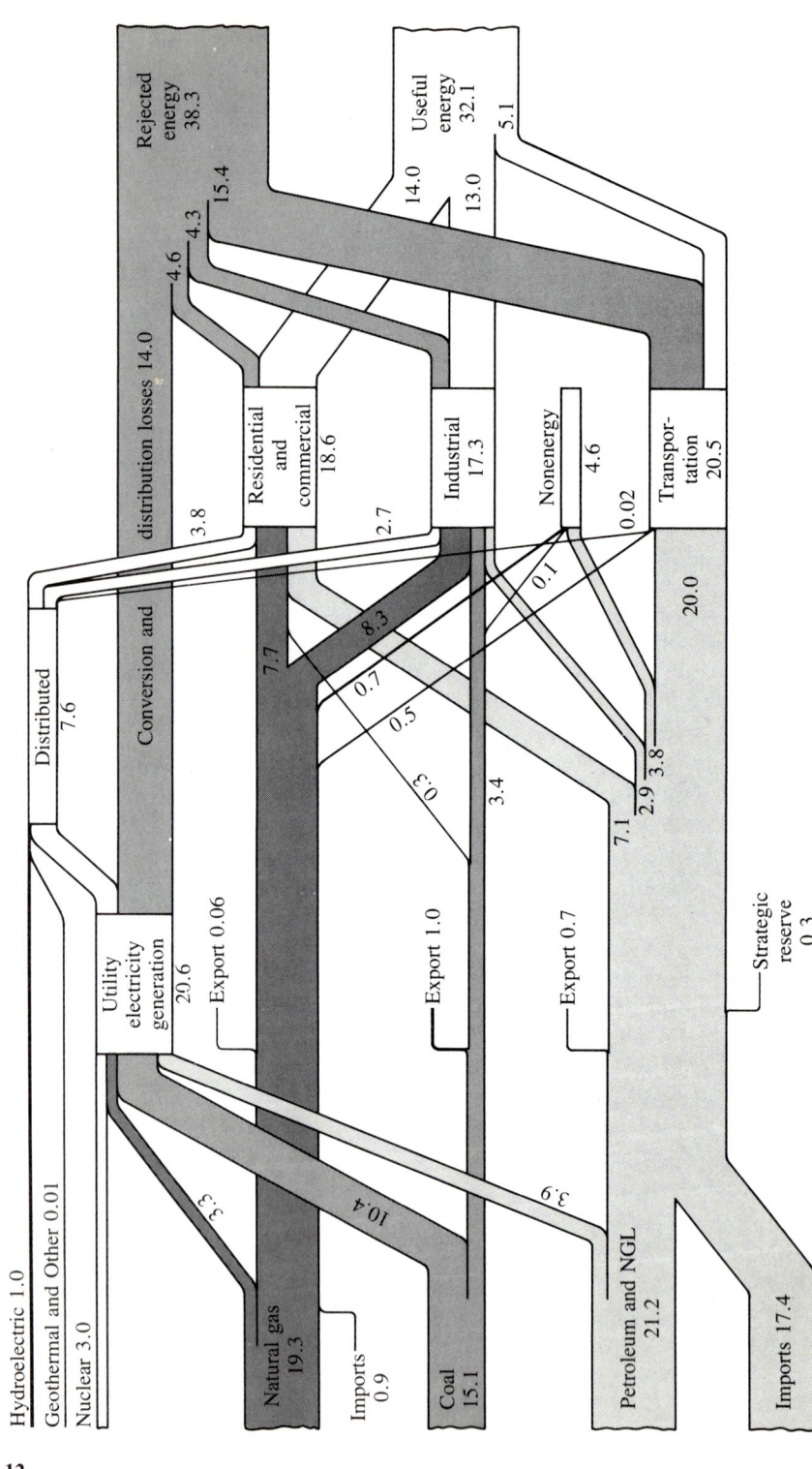

Figure 1.3 Energy flow in the United States in 1978 (*Ramsey, 1979*). Numbers are quads (quadrillions of Btu's).

Table 1.1 Solar cell development and production goals of the U.S. DOE National Photovoltaics Program

Costs per W_{pk} delivered with illumination at 1 kW/m² in 1980 U.S. dollars (DOE, 1980). System prices in third column include balance-of-system components required in addition to solar cells, whose prices appear in column 2.

Application and year	Collector price (FOB), $/W_{pk}$	System price, $/W_{pk}$	Production scale, MW_{pk}/yr	User energy price, ¢/kWh
Remote stand-alone (1982)	≤ 2.80	6–13		
Residential (1986)	≤ 0.70	1.60–2.20	100–1000	3.5–10.5
Intermediate load center (1986)	≤ 0.70	1.60–2.60	100–10,000	5.0–13.5
Central power station (1990)	0.15–0.40	1.10–1.30	500–2500	4.0–10.0

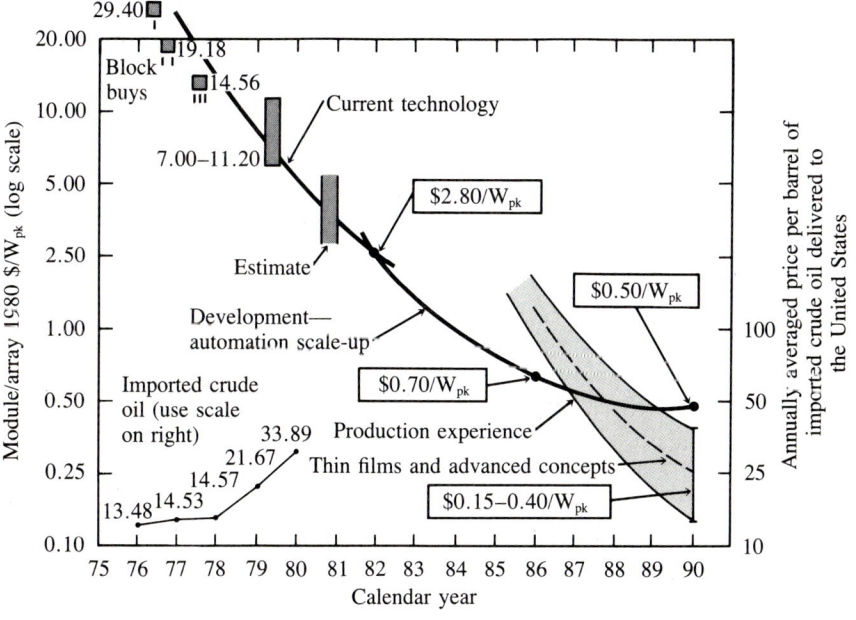

Figure 1.4 Price goals and cost history for terrestrial solar cell modules and arrays, in 1980 dollars. Shaded regions represent actual purchase prices. Block buys I, II, and III were large U.S. government purchases for demonstration projects. *(From DOE, 1980.)* The prices for a barrel of imported crude oil delivered at U.S. ports, averaged over suppliers and through the year, are also plotted in current year dollars. *(DOE, 1981)*

14 OVERVIEW

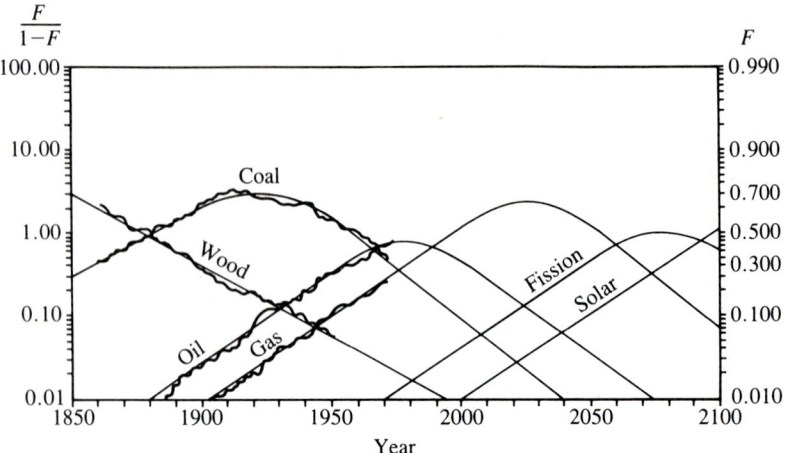

Figure 1.5 History and projection of energy sources for the world plotted versus year. F is the fraction of the total market supplied by a given energy source *(Weingart, 1978)*. The plot shows how similar the histories have been for all the major sources.

possible rates of expansion of manufacturing facilities, availability of materials, and so on. Encouraging answers have been obtained, as discussed in Chap. 4. The possible rate of growth of the solar energy supply is suggested by the curves in Fig. 1.5, where the growth in the market fraction commanded by a given energy source is plotted for both conventional and the newer alternative energy sources. If those logistic growth curves for wood, coal, natural gas, and oil also apply to solar sources, we should see a growth rate for all solar sources—solar cell and solar thermal power plants, and heating and cooling of buildings—that causes those sources to become dominant over a 40- to 50-year period.

Realizing that these predictions and trends are only suggestive of a possible future, let us begin our study.

REFERENCES

Barnett, A. A. (1978), *Physics Today*, Vol. 31, No. 3, 11.
DOE (1980), Photovoltaics Program, Subprograms, and Applications, U.S. Department of Energy Report 5210-5.
DOE (1981), *Monthly Energy Review*, U.S. Department of Energy Report DOE/EIA 0035(81/03), March, p. 74.
Ramsey, W. J. (1979), U.S. Energy Flow in 1978, Lawrence Livermore Laboratory Report UCID-18198, Livermore, California.
Weingart, J. M. (1978), The Helios Strategy, Int. Inst. for Appl. Sys. Anal. (IIASA) Report WP-78-8, Laxenburg, Austria.

PROBLEMS

1.1 *Energy payback time* The energy payback time is the time required for an energy-producing system to produce the amount of energy used in making the system. Suppose that by the introduction of new thin-film solar cells the energy required to make a solar cell system of given area is reduced by 50 percent from that required for a single-crystal system. Assume that the system's efficiency is also reduced, say, from 14 to 12 percent. What is the energy payback time of the new thin-film system if that for the single-crystal system is 6 years?

1.2 *Doubling time* Verify the results given in the box entitled "How Much Oil?" in Chap. 1. As a related exercise, suppose that you have a financial investment in which the amount of money increases by P percent each year. Show that the time in years for your money to double is approximately $70/P$.

1.3 *Efficiency of energy use* From Fig. 1.3 calculate the approximate percentage of total energy that is used in each of the major sectors listed below, together with the energy efficiency of each:

Sector	Approximate % of total energy input for sector	Approximate efficiency (%) of energy usage in sector
Utility electricity generation		
Industrial		
Residential and commercial		
Transportation		
Overall	(100%)	

CHAPTER
TWO
NATURE AND AVAILABILITY OF SOLAR ENERGY

CHAPTER OUTLINE

2.1 THE SUN AND THE SUN–EARTH RELATIVE MOTION
BOX: THE SUN AS A FUSION REACTOR
2.2 ATMOSPHERIC EFFECTS ON SOLAR RADIATION
2.3 SOLAR RADIATION MEASUREMENT AND INSTRUMENTATION
2.4 GEOGRAPHICAL DISTRIBUTION OF AVERAGE INSOLATION
2.5 EFFECTS OF COLLECTOR TILT
2.6 SUMMARY
REFERENCES
PROBLEMS

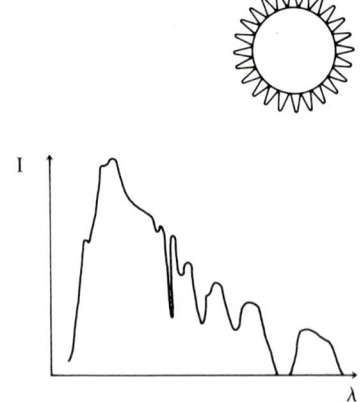

The sun, the energy source of all solar cell systems, is the subject of this chapter. A rather thorough description of the temporal and geographical variations of solar insolation is presented. It is hoped that, after studying this chapter, one can estimate the solar energy available to flat-plate and tracking concentrator collectors (solar cells) and the best ways of mounting them at a given location.

2.1 THE SUN AND THE SUN–EARTH RELATIVE MOTION

The sun is located in one of the spiral arms, the Orion arm, of our galaxy. It is believed that hydrogen nuclei are converted into helium nuclei under high temperature and pressure near the center of the sun by thermonuclear fusion. The fusion energy is transferred outward by radiation and convection and finally radiated into space in the form of electromagnetic waves in and near the visible spectrum. The spectrum of solar radiation is close to that of a blackbody heated to 5743 K.

A sun-centered view of the sun–earth system is shown in Fig. 2.1a. The earth's orbit is slightly elliptical and eccentric. The sun–earth distance varies

---

THE SUN AS A FUSION REACTOR

For thermonuclear fusion both high temperature and high concentration (density or pressure) of reacting gases are desirable. Near the center of the sun, the temperature is on the order of 10^7 K and the density (of hot gases!) is about 100 times that of water. The high density is maintained by the gravitational forces resulting from the sun's large mass. Current estimates of the power released from the sun are around 3.8×10^{26} W. At this rate, 2×10^{19} kg of hydrogen is consumed by fusion each year. This huge mass is only $1/10^{11}$ of the sun's weight.

To achieve controlled fusion as an energy source on earth in imitation of the sun, other means of confining the hot gases must be found. One method being investigated is called *magnetic* confinement. A low but sufficient gas pressure is maintained by confining the ionized gases in a "bottle" formed with a magnetic field. The other, newer method under study is called *inertial* confinement. A small, hollow glass bubble holding fusion gases is irradiated from several symmetrical directions with beams of intense laser light for less than one billionth of a second. The resultant implosion compresses the gases to a density comparable to that of lead for the brief duration of the fusion process.

Question: Is the gravitational force larger or smaller near the center of the sun than near its surface? How do you reconcile this with the fact that density (pressure) is larger near the center? (*Hint:* Force per unit volume is the gradient of pressure.)

18 NATURE AND AVAILABILITY OF SOLAR ENERGY

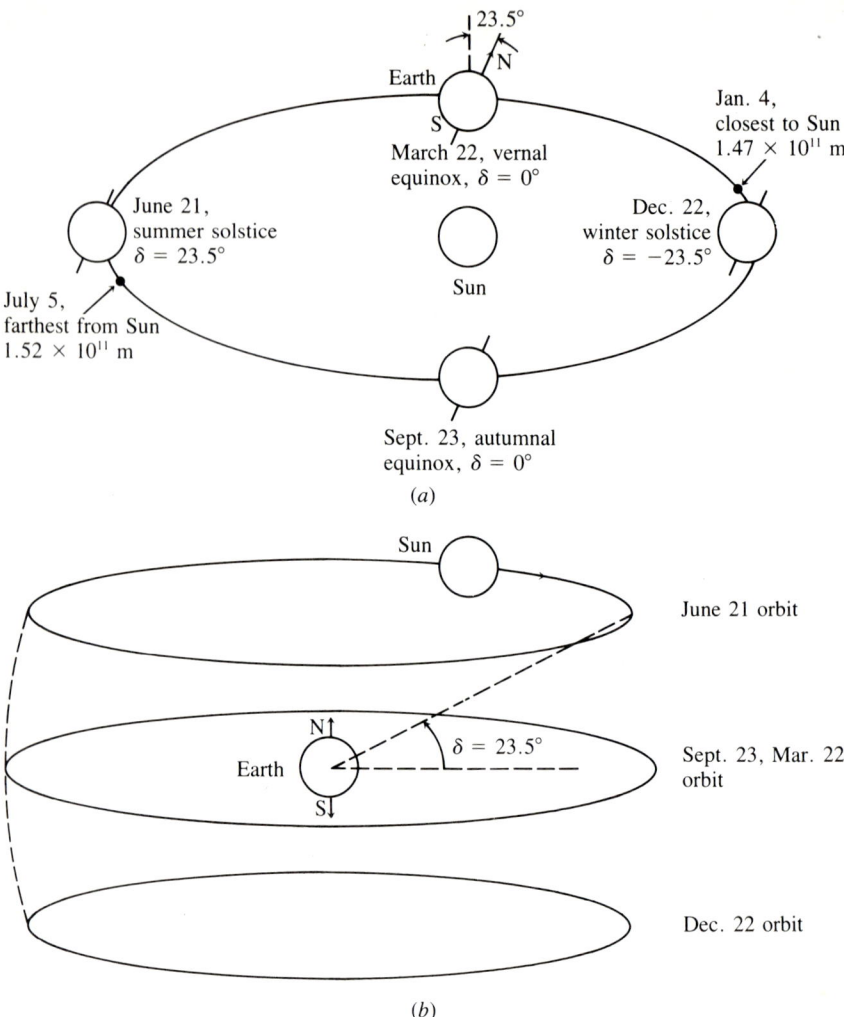

Figure 2.1 (a) The conventional sun-centered view of the sun-earth system. (b) An earth-centered view, which is easier to visualize. For example, the declination angle δ between the sun ray and the plane of the equator is better illustrated in b. The date given may vary by one day or so.

seasonally about the mean by 1.7 percent. At the mean distance of 1.495×10^{11} m, known as one *astronomical unit*, the solar flux outside the earth's atmosphere is 1.353 kW/m², a quantity known as the *solar constant*. The earth spins around its own polar axis once every space day. Its polar axis is tilted by 23.45° from the normal of the orbital plane. This causes seasonal variations of the sunlight incidence angle.

The sun–earth relationship can be visualized more easily in an earth-centered view, which is shown in Fig. 2.1b. The two views, of course, are equivalent through coordinate transformations. In Fig. 2.1b, the earth is motionless with its polar axis pointing upward. The sun moves around the earth exactly once every day at constant angular speed (15° per hour) tracing an almost perfectly centered and circular path. The solar path is highest over the north pole with the sun rays making a 23.5° angle with the plane of the equator around June 21. Around September 23 and March 22, the sun rays arrive parallel to the plane of the equator; around December 22, the solar path is lowest below the south pole, making a $-23.5°$ angle with the plane of the equator. On the nth day of the year, the declination δ can be found from the approximate equation

$$\delta = 23.45° \sin\left[2\pi \cdot \frac{n - 80}{365}\right] \qquad (2.1.1)$$

The radii of the solar paths also vary through the year so that the earth–sun distance varies as in Fig. 2.1a, or approximately

$$\text{Earth-sun distance} = 1.5 \times 10^{11}\left[1 + 0.017 \sin\left(2\pi \frac{n - 93}{365}\right)\right] \text{ m} \qquad (2.1.2)$$

To visualize the geometrical factors of solar flux at a certain location on earth, imagine a tangent plane touching the earth at that location, as shown in Fig. 2.2. The plane is an extension of the earth's horizon at that point. The angle between this plane and the polar axis is equal to the latitude. The sun is visible to an observer at that location only when the sun's path is above the plane. Point A corresponds to sunrise and point B, sunset. The midpoint between A and B corresponds to the "solar noon." Clearly, for any location in the northern hemisphere, days are longer in June than in December. Moreover, a horizontal collector (such as the earth's surface) is closer to being normal to the solar flux in June than in December. Therefore, June is warmer than December. To enhance solar energy collection in the winter months, one can tilt the collector surface toward the south. It can be shown that the total annual collection is maximized when the collector is approximately parallel to the polar axis, i. e., the collector is tilted to the south by ϕ (see Sec. 2.5). The following are examples of the many questions that can be answered using Figs. 2.1b and 2.2.

Imagine a tracking solar collector in front of you. Can you visualize the motions it must go through in order to face the sun directly at all daytime hours throughout a year?

From the symmetry in Fig. 2.2 (the earth's radius is negligible in comparison with the sun–earth distance), can you see that the annual average length

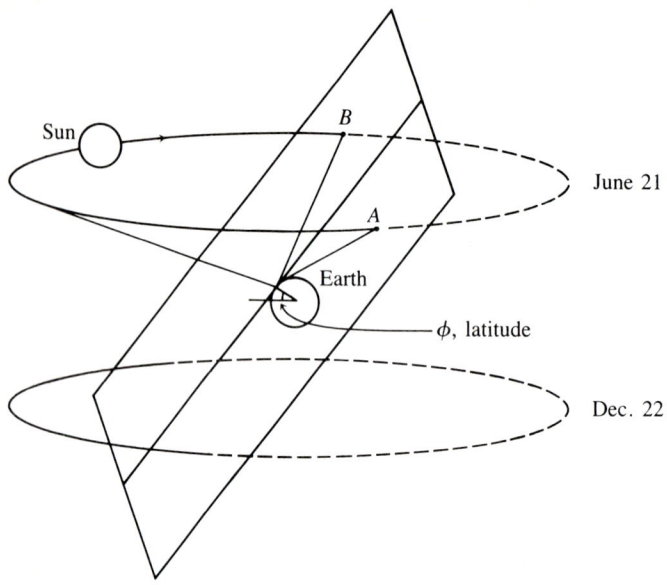

Figure 2.2 A tangent plane touching the earth is but an enlargement of the horizontal plane at that location. This plane can help one visualize the seasonal variation of the length of the day, the motions a tracking solar collector must go through in a day and a year, and so forth.

    of day is equal to the average length of night? If so, what is the average length of day? (Answer: 12 hours.)

Around the polar regions, the sun may be invisible for an entire day. In which portions of the world will this phenomenon occur on at least one day each year? (Answer: north of 66.5°N and south of 66.5°S.)

2.2 ATMOSPHERIC EFFECTS ON SOLAR RADIATION

On a clear day and when the sun is directly overhead, 70 percent of the solar radiation incident to the earth's atmosphere reaches the earth's surface undisturbed. Another 7 percent or so reaches the ground in an approximately isotropic manner after scattering from atmospheric molecules and particulates. The rest is absorbed or scattered back into space. Figure 2.3 illustrates the situation. (For a review of the various processes see DOE Report HCP/T2552-01.) Clearly, both the direct and scattered fluxes vary with time and location because the amounts of dust and water vapor in the atmosphere are not constant even on clear days. A commonly accepted set of solar fluxes

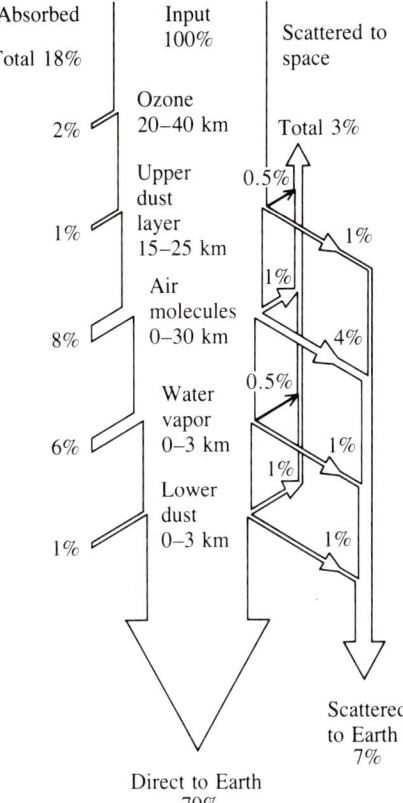

Figure 2.3 Typical air mass one (AM1) clear sky absorption and scattering of incident solar energy. Clouds can significantly reduce the direct transmission and increase scattering.

for air mass one (AM1) sun is shown in Table 2.1. A more complete table of units and numerical constants can be found in Appendix 2.

Air mass one refers to the thickness of the atmosphere a sunbeam passes through if the beam is normal to the horizon. The angle between the sunbeam and the horizon, α, is called the *solar altitude*. For any solar altitudes other than 90°, the air mass number is of course larger and is equal to csc α except for very small α's, for which the curvature of the atmosphere makes the air mass smaller than csc α. Since an AM1 atmosphere reduces the direct flux by the factor 0.7 (see Table 2.1), one would expect the direct intensity at an arbitrary air mass number to be

$$I = 1.353 \times 0.7^{\text{air mass}} = 1.353 \times 0.7^{\csc \alpha} \text{ kW/m}^2 \quad (2.2.1)$$

Meinel and Meinel (1976) have found, however, that actual observations by Laue (1970) can be fitted better with

$$I = 1.353 \times 0.7^{(\csc \alpha)^{0.678}} \text{ kW/m}^2 \quad (2.2.2)$$

Table 2.1 Solar fluxes of AM1 sun

Sunbeam normal to horizon, as at solar noon on summer solstice at 23.5°N latitude [Meinel and Meinel (1976)]

Above atmosphere (direct/total):	1.353 kW/m^2
Desert sea level (direct):	0.970 kW/m^2
Desert sea level (total):	1.050 kW/m^2
Standard sea level (direct):	0.930 kW/m^2

$$1 \text{ kW/m}^2 = 1.433 \text{ langley/min (cal/cm}^2 \cdot \text{min)}$$
$$= 316.9 \text{ Btu/ft}^2 \cdot \text{hr}$$

There is no known explanation for the additional exponent. Equations (2.2.1) and (2.2.2) can be used to estimate the clear sky direct normal (collector facing sun) flux at arbitrary solar altitude.

Besides intensity, the spectral distributions of solar fluxes are also affected by the atmosphere. This fact is important to photovoltaic solar energy conversion since the conversion efficiencies of solar cells depend on the spectrum of the incident light. Figure 2.4 shows the solar spectrum outside the atmosphere (air mass zero, or AM0), which is close to the 5743 K blackbody radiation spectrum, and the AM1 (air mass one) direct and estimated diffuse radiation spectra at sea level. The AM1 direct radiation spectrum is from Thekaekara (1974). Absorption by ozone is essentially complete below 0.3 μm wavelength. The relatively large attenuation below 0.8 μm is due to scattering by molecules and particulates. These scattering processes become weaker at longer wavelengths, as has been shown by both theory and observation. This also explains the spectrum of the diffuse radiation, which is richer than the direct radiation in the blue portion of the spectrum. The many notches in the sea-level spectrum can be attributed to the absorption bands of various atmospheric gases. Some of the attributions are shown in Fig. 2.4.

From the AM0 and AM1 spectra, one may estimate the spectrum for an arbitrary air mass AMK by assuming that the attenuation at each wavelength follows the form of Eq. (2.2.1) or Eq. (2.2.2). For example,

$$I_{\text{AM}K}(\lambda) = I_{\text{AM0}}(\lambda) \left| \frac{I_{\text{AM1}}(\lambda)}{I_{\text{AM0}}(\lambda)} \right|^{K^{0.678}} \quad (2.2.3)$$

In analyzing the performance of solar cell systems, the cell output is usually assumed to be proportional to the solar radiation intensity with no regard to the variations in the spectral distributions. This practice should be satisfactory for engineering purposes.

So far only the clear sky condition has been considered. The amount of sky cover in the form of clouds is a dominant factor in determining the transmission and scattering of solar radiation. Although the amount of water

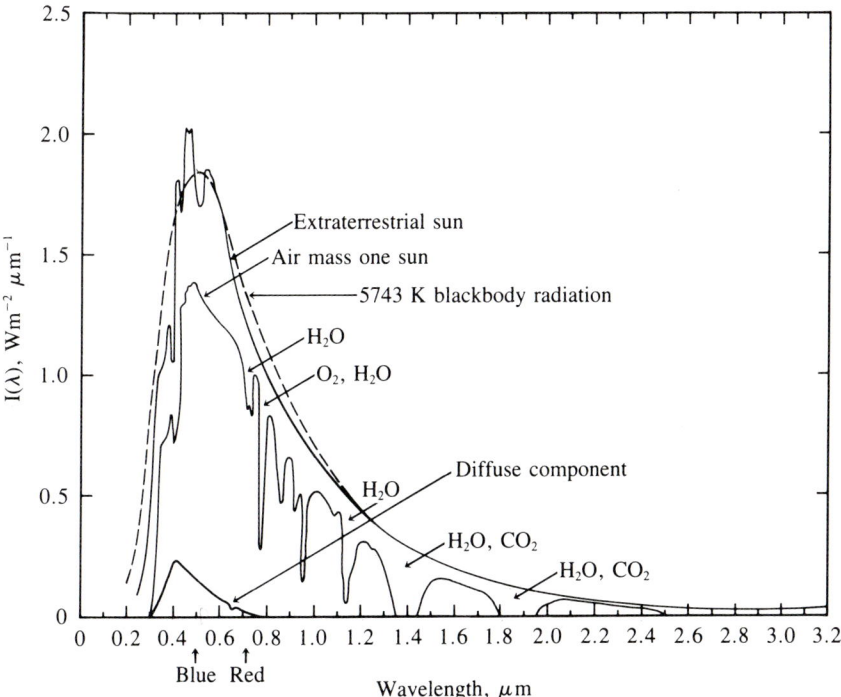

Figure 2.4 Spectral intensities of the extraterrestrial and air mass one (sun directly overhead, measured at sea level) direct solar radiation. A typical spectrum of clear sky scattered radiation is also shown.

in the clouds is usually a small fraction of the total water content in the air, the condensed droplets or ice crystals have much stronger effects on light than does water vapor. Unlike the sun–earth geometry and the air mass number, the amount of sky cover defies simple modeling. As a result, insolation may fluctuate unpredictably over intervals of minutes. Even longer-term average insolation can be obtained only through measurements, as discussed in the next two sections.

2.3 SOLAR RADIATION MEASUREMENT AND INSTRUMENTATION

Many different types of instruments are used for measuring solar radiation, including the following:

Pyranometer: Total (direct and diffuse) radiation
Shading-ring pyranometer: Diffuse radiation

Moving shadow-bar pyranometer: Both total and diffuse radiation
Pyrheliometer: Direct radiation at normal incidence
Sunshine recorder: Hours of bright sunshine

Pyranometer Also known as a *solarimeter*, the pyranometer is generally mounted in a horizontal position away from tall objects so that the 2π field of view of the instrument covers the entire sky. It responds equally to the energy in all wavelengths. The Eppley pyranometer (see Fig. 2.5a) is the type most commonly used in the United States. It employs metal wedges arranged into a circular disk and alternately painted with Parson's black for black and magnesium oxide for white. The disk is protected by one or two layers of dome-shaped glass covers. The temperature difference between the black and white wedges is sensed by multiple thermocouple junctions whose output voltage is measured. Similar instruments are manufactured in Europe as Kipp and Zonen pyranometers.

Because of the ease of operating pyranometers, the vast majority of solar insolation data is gathered with this type of instrument. In some parts of the world, the data are mainly gathered with a simpler kind of instrument based on the differential expansion of bimetal elements. The principles and the calibration characteristics of many pyranometers are discussed by Robinson (1966).

Shading-ring pyranometer A ring-shaped hoop sunshield may be added to a pyranometer to exclude direct sunlight and thereby permit measurement of the diffuse components. When this reading is subtracted from that of a standard pyranometer, the result is the direct solar radiation. To keep the obstruction of the sky small, the ring is made narrow, shading only about 5°, and the position of the ring is changed every few days (see Fig. 2.5b).

Moving shadow-bar pyranometer This instrument is a clever combination of the standard pyranometer and the shading-ring pyranometer. Instead of the quasi-stationary shading ring, a narrow shadow bar is moved over the sensor every few minutes, shading the direct radiation and causing a drop in the recorder trace. The upper envelope of the trace thus provides a record of the total radiation versus time, and the lower envelope is a record of the diffuse radiation. The sensor is a small silicon diode (solar cell) having a short response time.

Pyrheliometer The pyrheliometer has a small field of view, around 6°, and tracks the sun continuously and thus measures the direct normal radiation. Since the field of view is larger than the 0.53° subtended by the solar disk, the reading is higher than the true direct flux by a few percent. Pyrheliometer stations are relatively rare. For example, among the approximately 100 sta-

2.3 SOLAR RADIATION MEASUREMENT AND INSTRUMENTATION

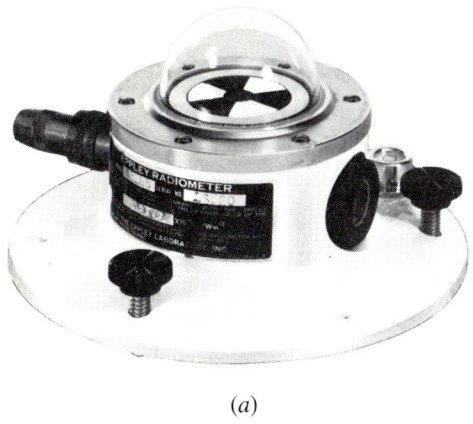

(a)

(b)

Figure 2.5 (a) Pyranometer. (b) Pyranometer with shading ring. (*Courtesy Eppley Laboratory, Inc.*)

26 NATURE AND AVAILABILITY OF SOLAR ENERGY

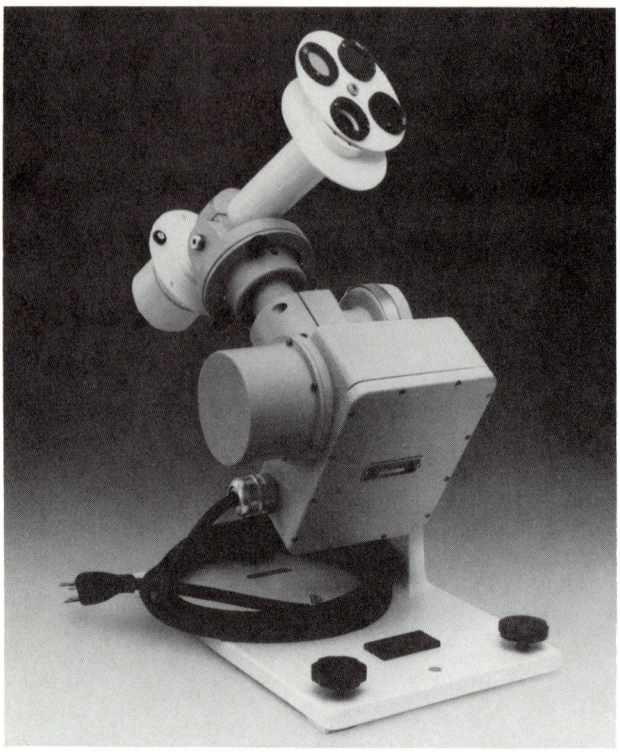

(c)

Figure 2.5 *(continued)* (c) Pyrheliometer. (*Courtesy Eppley Laboratory, Inc.*)

tions that record solar radiation in the United States, only about 18 record direct normal insolation. Pyrheliometer measurements are needed to predict the performance of tracking concentrator photovoltaic systems. Figure 2.5c shows a pyrheliometer with filters.

Sunshine recorder The sunshine recorder measures the duration of "bright sunshine," the number of hours per day that the sunlight intensity is above some chosen level. An older instrument, the Campbell-Stokes recorder, uses a spherical lens that focuses the sunlight on an advancing treated paper. The paper is burned whenever the beam radiation is above a critical level. This type of recorder is used in many hundreds of stations throughout the world. The standard U.S. Weather Bureau instrument uses two photocells, one of which is shaded from the direct radiation. When the reading difference between the two cells is above a set threshold, "bright sunshine" is assumed to be present. Although sunshine-recorder data does not provide direct information on the incident solar energy, it is widely available and may be used to estimate the solar radiation (Duffie and Beckman, 1974).

2.4 GEOGRAPHICAL DISTRIBUTION OF AVERAGE INSOLATION

In designing a photovoltaic system the temporal variability of solar radiation needs to be considered for the following reasons.

Daily and hourly variations Daily variations are important for the design of storage capacity in totally self-sufficient systems. The variations from hour to hour, or even from minute to minute, should be considered when accurately calculating the system output since the output power of the cells and the system in general is not linearly proportional to the solar radiation. For example, one hour of 1 kW/m^2 radiation and three hours of zero radiation generally would not produce the same electrical energy as four hours of 0.25 kW/m^2 radiation.

Unfortunately, variations on such fine time scales are difficult to record, present, and adequately deal with in design. The data of these variations will not be presented here. Statistical techniques will no doubt be used in this area more extensively in the future. Interested readers might note that the U.S. National Oceanic and Atmospheric Administration (Asheville, North Carolina) keeps on tape or punched cards the daily and (for some 40 stations) hourly radiation data recorded at many locations for varying periods, starting from 1952.

Monthly variations Monthly variations result from the seasonal changes of both the sun–earth relationships (Figs. 2.1 and 2.2) and weather, particularly cloudiness. With the average monthly sunshine data, one may estimate the monthly outputs of a photovoltaic system for comparison with the expected monthly electricity demands.

Yearly variations The yearly average insolation is normally used to analyze the average energy cost, the energy payback time (defined in Chap. 4), and other such information about a system. At a given location, the average insolation varies from the mean from year to year by less than 10 percent. There is no clear pattern (cycles) in the year-to-year variations. However, after major volcanic eruptions, large temporary decreases in yearly insolation have been observed.

Figure 2.6 shows the average insolation by month at selected sites in the United States. The averages are for periods of many years and are expressed in terms of kWh/m$^2 \cdot$ day. Both direct normal radiation (excluding diffuse radiation, collector normal to sunbeams at all times), corresponding to the flux on tracking concentrators, and the radiation on horizontal surfaces are shown. The monthly variations can be explained by Fig. 2.7. In the northern hemisphere, the radiation on a horizontal surface generally peaks in June and reaches a minimum in December because of the seasonal changes in the

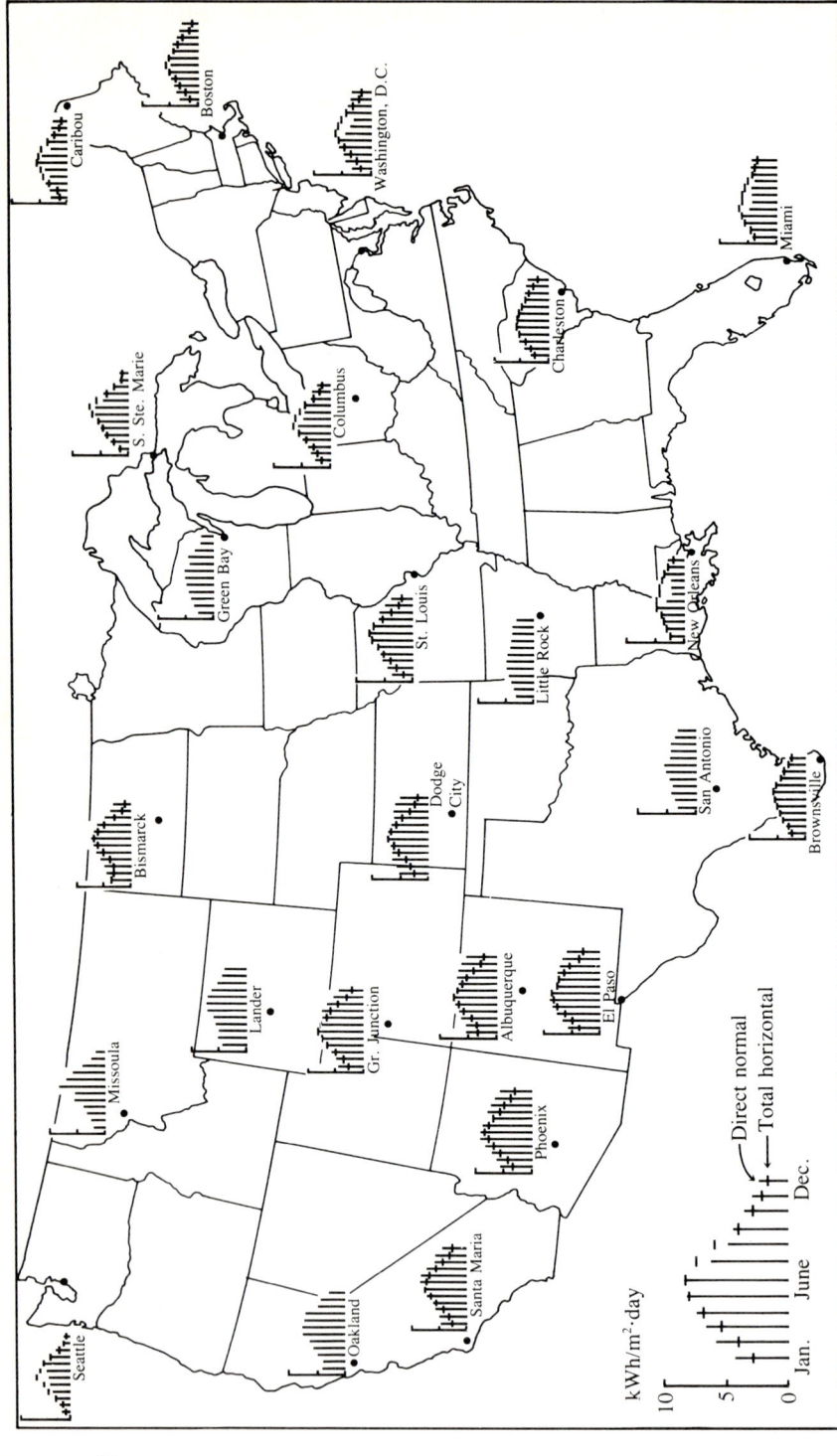

Figure 2.6 Daily direct normal and total (direct plus diffuse) horizontal radiation in selected United States cities. *(DOE Report HCP/T2252-01, 1978, and HCP/T-4016/1, 1979).* Total horizontal insolation data are not available at some of the locations shown.

2.4 GEOGRAPHICAL DISTRIBUTION OF AVERAGE INSOLATION

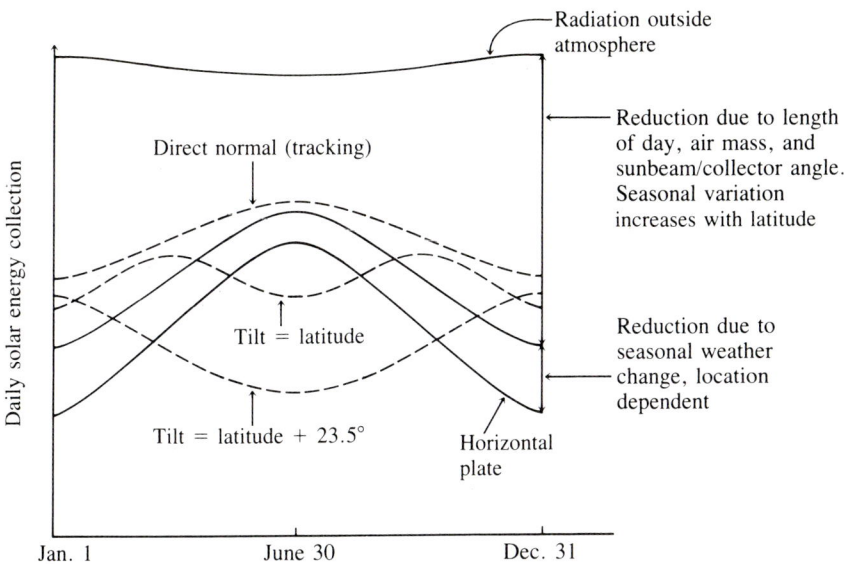

Figure 2.7 Qualitative plots of seasonal variations of daily solar energy collection for a location north of 23.5°N. "Tilt" (see Sec. 2.5) is the angle between the collector plate and the horizontal plane. The plate is tilted due south.

length of the day (Fig. 2.2), the average air mass, and the average angle between sunbeam and the horizontal surface (collector). The peak-to-valley ratio increases with increasing latitude. At the same latitude, the insolation tends to be lower in the eastern regions of the United States because of the weather conditions. The average direct normal radiation is usually greater than the radiation on horizontal surfaces because the angle between sunbeam and collector is always 90° in the case of direct normal radiation. The west–east difference is even clearer with the direct normal radiation because the direct normal flux is more sensitive to cloudiness.

Figures 2.8 and 2.9 show contour drawings of annual radiation on a horizontal surface and the direct normal radiation in the United States, respectively, in units of $kWh/m^2 \cdot day$. It is worth remembering that the average energy received on a horizontal surface per day is between 3.5 and 6 kWh/m^2, or the equivalent of 4 to 6 hours of the clear-sky noontime sun every day of the year. The energy received by a tracking collector ranges between 3.5 and 7.5 kWh/m^2.

Figures 2.10a through d are world maps of energy received daily on a horizontal surface in four months of the year. Notice the high insolation in the desert zones around 30°N and 30°S where the sky coverages are statistically low. Table 2.2 shows the daily energy density averaged over a month and a year at selected cities.

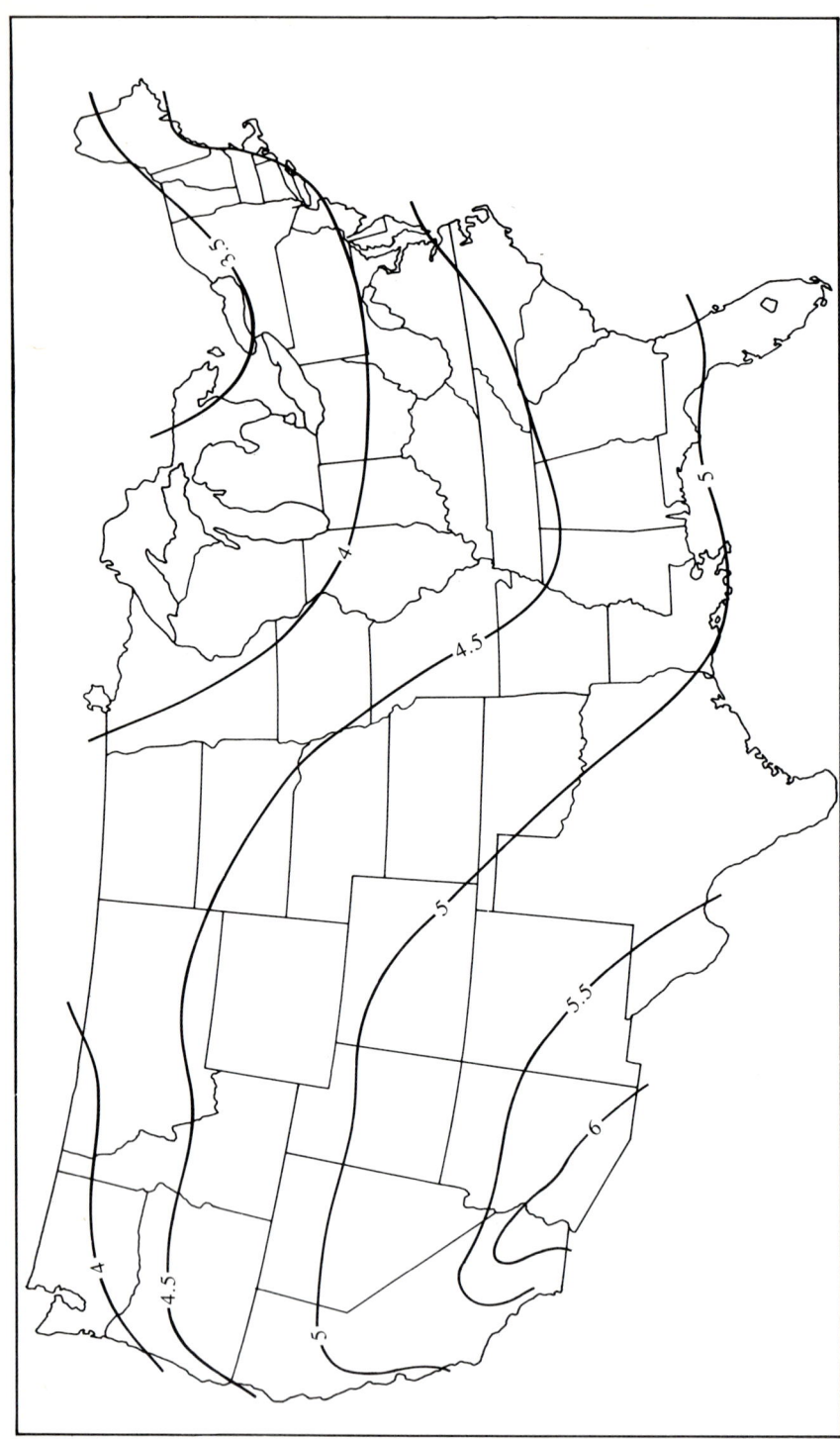

Figure 2.8 Contour map of daily radiation (direct plus diffuse) on a horizontal surface in the United States. The values are averaged over many years and expressed in kWh/m^2 · day. *(From DOE Report HCP/T2252-01.)*

Figure 2.9 Contour map of daily radiation (direct only) on a tracking collector facing the sun at all times in the United States. The values are averaged over many years and expressed in kWh/m^2 · day. *(From DOE Report HCP/T2252-01.)*

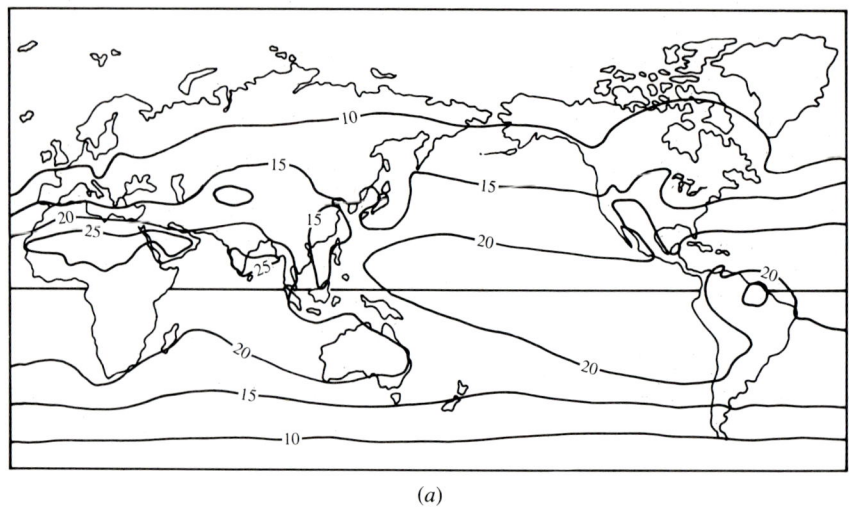

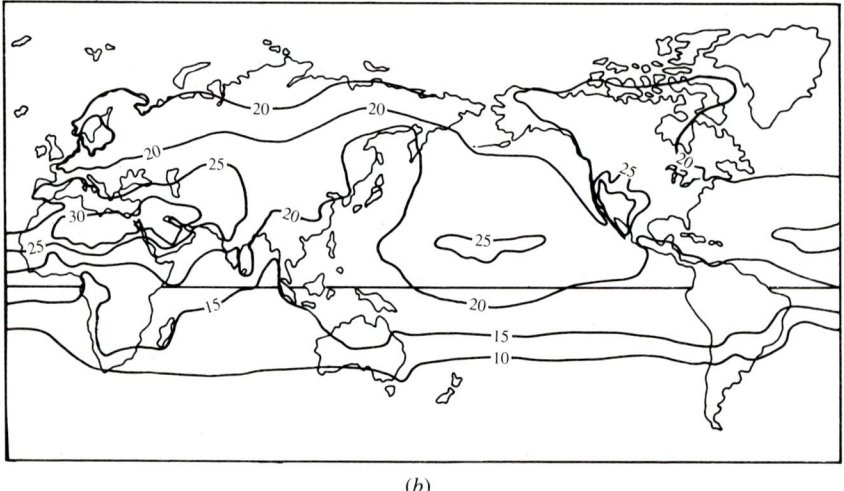

Figure 2.10 Global isoflux contours. Total insolation in MJ/m^2 · day on a horizontal surface in (*a*) March, (*b*) June, (*c*) September, and (*d*) December. 1 MJ/m^2 = 0.278 kWh/m^2. *(Meinel, 1978.)*

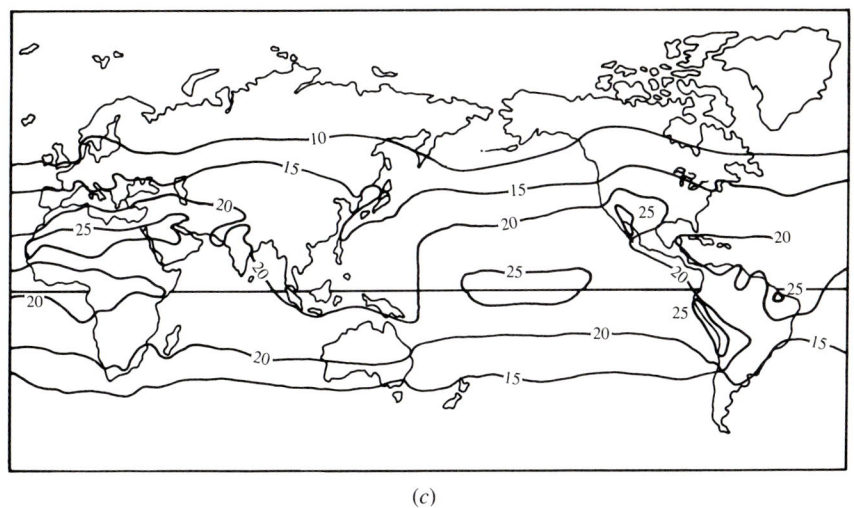

(c)

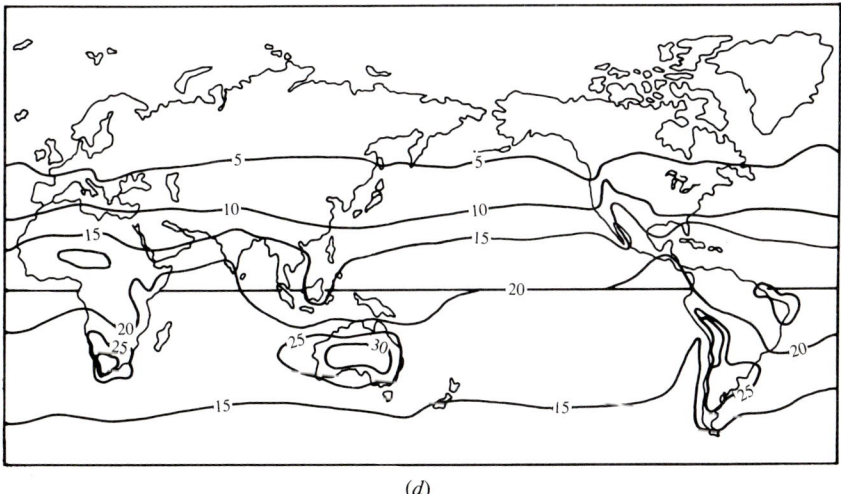

(d)

Figure 2.10 *(continued)*

Table 2.2 Daily direct and diffuse radiation energy on a horizontal surface in kWh/m² · day

The direct normal radiation energy in Albuquerque is also shown on the second line of entries for that city. [Meinel and Meinel (1976) and DOE Rept. HCP/T2552–01.]

Location	Latitude, degrees N	Jan	Feb	Mar	Apr	May	June	July	Aug	Sept	Oct	Nov	Dec	Yearly average
Yangami, Congo	01	4.76	5.24	5.33	5.19	5.10	4.62	4.10	4.20	4.75	4.77	4.92	4.34	4.78
Dakar, Senegal	15	5.35	6.26	7.37	7.29	7.20	6.75	5.96	5.30	5.40	5.22	5.26	5.47	6.07
Calcutta, India	22	6.19	7.18	8.15	9.08	9.12	9.50	9.49	9.31	7.50	7.26	6.48	5.83	7.92
Honolulu, U.S.	22	4.22	4.91	6.00	6.50	7.18	7.15	7.15	7.12	6.66	5.90	4.95	4.31	6.00
Tucson, U.S.	32.5	3.66	4.55	5.28	7.61	8.47	8.12	7.28	6.88	6.63	5.14	4.14	3.54	5.94
Albuquerque, U.S.	35	3.57	4.24	5.78	7.05	7.86	8.67	7.89	7.25	6.36	5.40	4.05	3.42	5.96
		6.26	6.05	5.76	8.09	7.92	8.54	7.90	6.84	7.43	7.69	4.75	5.37	6.88
Tokyo, Japan	36	2.21	2.68	3.18	3.63	3.98	3.52	3.91	3.93	2.95	2.35	2.15	1.96	3.04
New York, U.S.	42	1.42	2.22	3.01	4.22	4.96	5.19	5.11	4.26	3.68	2.83	1.72	1.25	3.32
Brussels, Belgium	51	0.65	1.25	2.39	4.02	4.72	5.13	4.72	4.11	2.92	1.83	0.88	0.55	2.76

2.5 EFFECT OF COLLECTOR TILT

Except for locations near the equator, laying a flat photovoltaic-cell panel on a horizontal surface is not the best design. Tilting the plate toward the equator, i.e., toward the south for a northern hemisphere location, can increase the total annual solar energy collection and smooth out the difference between summer and winter collections.

By referring to Figs. 2.1b and 2.2, one can see that a horizontal collector at latitudes north of or near 23.5°N would collect the maximum daily energy on June 21 and the minimum on December 22 (as shown in Fig. 2.7). If the collector makes an angle with the horizontal plane equal to the latitude ϕ (see Fig. 2.2 and the inset of Fig. 2.11), the collector would be normal to the sun's orbit on September 23 and March 22 and the maximum energy collection would occur near these two dates (as shown in Fig. 2.7). Thus the seasonal variations in energy collection are reduced by tilting. If one tilted the collector by an additional 23.5°, one could maximize the energy collection on December 22. This may be desirable for a solar heating system, but usually offers no advantages for a photovoltaic system.

Most available insolation data are for horizontal collectors. The unknown directional distributions of the diffuse radiation make it impossible to calculate accurately energy collection by a tilted plate from the measured collection by a horizontal surface. Neglecting seasonal variations of weather, Morse and Czannecki (1958) calculated the relative annual insolation due to direct radiation only as a function of the tilt angle, s. They concluded that the maximum annual energy collection per collector area is achieved when $s \approx 0.9\phi$, as shown in Fig. 2.11. With s slightly less than ϕ, the plate faces the sun more directly in the summer, when days are longer, than in the winter, and the

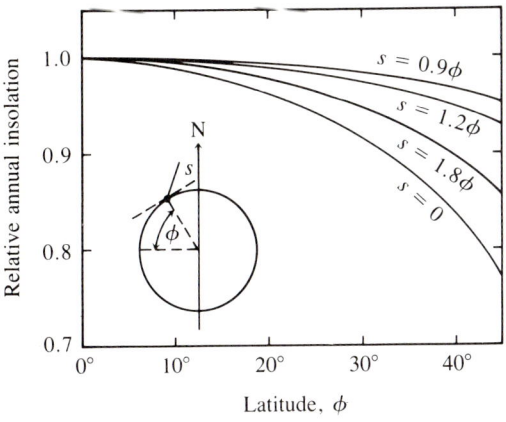

Figure 2.11 Relative annual insolation due to direct radiation on surfaces tilted at various degrees toward the equator. *(From Morse and Czarnecki, 1958.)*

annual collection is maximized. To estimate the insolation on a tilted collector, one can look up the annual insolation on a horizontal surface from the figures in Sec. 2.4 and multiply it by the ratio of the annual collection by a tilted plate to the collection by a horizontal plate ($s = 0$) obtained from Fig. 2.11. From the symmetry in Fig. 2.2, one can conclude that the collector should be tilted toward due south. However, Morse and Czannecki (1958) found that even a 22.5° tilt toward east or west causes less than 2 percent reduction in annual insolation for latitudes up to 45°.

Yet another alternative is to change the tilt angle s seasonally so that the envelope of the curves for fixed tilt angles in Fig. 2.7 is achieved. The annual insolation in this case would lie between the best fixed-tilt insolation and the direct normal insolation, which may be found from Fig. 2.9.

2.6 SUMMARY

The geometrical relationship between the earth and sun (Figs. 2.1*b* and 2.2) can help us understand the seasonal changes of insolation and the effect of tilting the solar collector and also to visualize the motions that a tracking mechanism must produce. The unpredictability of weather, particularly the cloud coverage of the sky, necessitates that measured insolation data be used for predicting and analyzing the performance of photovoltaic systems. The average (over a month and a year) daily collections of solar energy by a horizontal or tracking collector of unit area are presented in Figs. 2.6 through 2.10. Tilting the collector toward the equator to just short of paralleling the polar axis maximizes the total annual energy collection. In analyzing a photovoltaic system, it is usually assumed that its total or average electrical energy output is proportional to the total or average insolation, even though the electrical output actually varies with the solar intensity nonlinearly and also depends on the spectrum of the solar radiation (see Sec. 3.6).

REFERENCES

DOE (1978), On the Nature and Distribution of Solar Radiation, Report HCP/T2552-01, prepared by Watt Engineering Ltd. Available from U.S. Government Printing Office.
DOE (1979), Solar-Climatic Statistical Study, Report HCP/T-4016/1, prepared by Northrop Services, Inc. Available from National Technical Information Service.
Duffie, J. A., and Beckman, W. A. (1974), *Solar Energy Thermal Processes*, Wiley-Interscience.
Laue, E. G. (1970), *Sol. Energy, 13,* 1, 43.
Meinel, A. B., and Meinel M. P. (1976), *Applied Solar Energy,* Addison-Wesley Publishing Company.

Morse, R. N., and Czarnecki, J. T. (1958), Report E.E. 6 of Engineering Section, Commonwealth Scientific and Industrial Res. Organization, Melbourne, Australia.

Robinson, N. (1966), *Solar Radiation,* Elsevier, London.

Thekaekara, M. P. (1974), Supplement to the Proc. 20th Annual Meeting of Inst. for Environmental Sci., p. 21.

PROBLEMS

2-1 *Sun-earth relative motion* Answer the three questions at the end of Sec. 2.1. Give your reasoning.

2-2 *Atmospheric attenuation of solar radiation*
  (*a*) Explain the rationale of Eq. (2.2.1).
  (*b*) Sketch the approximate AM2 solar spectrum from the AM0 and AM1 spectra in Fig. 2.4 and Eq. (2.2.3).

2-3 *Shading-ring pyranometer* Based on your understanding, sketch the shading-ring arrangement of a shading-ring pyranometer. How should the ring be positioned relative to the earth's coordinates? How should it be adjusted every few days?

2-4 *Seasonal variation of solar insolation* Redraw Fig. 2.7 for a location at 23.5°S.

2-5 *Tilted collector* What needs to be known or what model information is necessary to convert pyranometer solar data to the flux on a tilted plane?

2-6 *Solar insolation in your location* What is the average daily total horizontal radiation in your location? Estimate the daily direct normal radiation, the total radiation on a flat plate tilted to maximize the annual collection, and the month-by-month changes of all these quantities. Where would you go for more information? If you were to supply about one third of the electricity use of a typical household with a 10 percent efficient photovoltaic system, how would you mount the solar cell panels and how large should they be?

CHAPTER

THREE

PRINCIPLES OF SOLAR CELL OPERATION

CHAPTER OUTLINE

3.1 ELEMENTS OF SOLAR CELL OPERATION
3.2 SEMICONDUCTORS
3.3 LIGHT ABSORPTION AND CARRIER GENERATION
3.4 CARRIER RECOMBINATION
3.5 *pn* JUNCTIONS
3.6 SHORT-CIRCUIT CURRENT
BOX: LIGHT SENSORS: PHOTODIODES AND PHOTOCONDUCTORS
3.7 EFFICIENCY
3.8 FACTORS AFFECTING THE CONVERSION EFFICIENCY
3.9 SUMMARY
REFERENCES
PROBLEMS

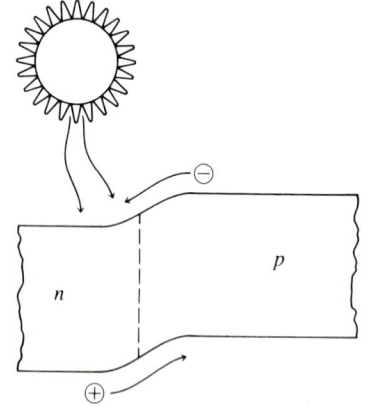

In this chapter we shall explore the principles of operation, the electrical characteristics, and the limits of efficiency of solar cells. Silicon *pn*-junction solar cells will often be used as examples, but the principles presented here are general. The choices of cell materials and structures will be discussed in more detail in later chapters.

Mathematics is kept to a minimum in this chapter. The occasional long equations are included for reference purposes and may be bypassed by the reader. It is hoped that a conceptual, physical, and intuitive understanding of the operation of solar cells will emerge from studying this chapter. For more rigorous, mathematical, and formal treatments of this subject, readers should consult technical journals or other books, for example, Hovel (1975).

3.1 ELEMENTS OF SOLAR CELL OPERATION

Figure 3.1 illustrates the basic operation of a solar cell. Light photons are absorbed by the semiconductor and in the process generate electric-charge carriers called *electrons* and *holes*. Electrons and holes diffuse to a "junction," either a *pn* junction as illustrated in Fig. 3.1 or some other type of junction where a strong internal electric field exists. Electrons and holes are separated by the field and give rise to electric voltage and current in the external circuit. This scenario involves several topics that will now be discussed individually: semiconductors, light absorption and carrier generation, *pn* junction, cell current, cell voltage, and efficiency of energy conversion.

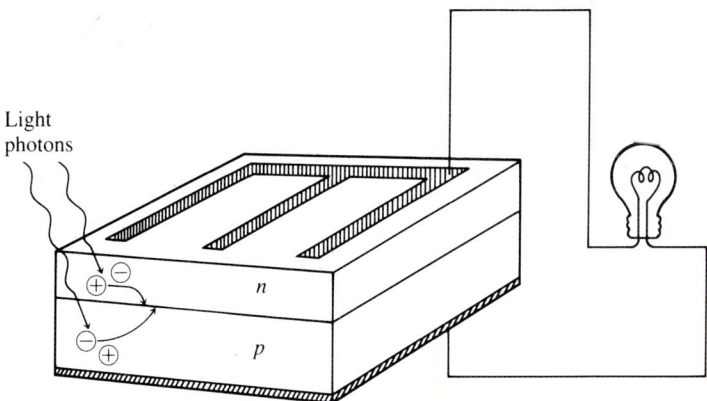

Figure 3.1 Positive and negative charge carriers are generated in a semiconductor as light photons are absorbed. These carriers, when collected by a *pn* junction, give rise to an electric current in the external circuit. Here, the current is shown to power a light bulb.

3.2 SEMICONDUCTORS

Good electrical conductors such as copper have conductivities in the range of 10^6 mho/cm (if 1 volt is applied to two opposing faces of a 1 cm × 1 cm × 1 cm copper cube, 10^6 amperes of current will flow between the faces). At the other extreme, good insulators such as quartz (SiO$_2$) have conductivities in the range of 10^{-16} mho/cm. Semiconductors have conductivities between these extremes in the range of 10^{-4} to 10^4 mho/cm. Furthermore, the conductivities of semiconductors may be changed within this range by adding small amounts of impurities, known as *dopants,* into the semiconductors. The conductivities of sufficiently pure semiconductors increase rapidly with rising temperature. These are the most easily identified traits of semiconductors.

Semiconductors may be elements such as silicon and germanium, or compounds such as cadmium sulfide (CdS) and gallium arsenide (GaAs), or alloys such as Ga$_x$Al$_{1-x}$As where x may be any value between zero and one. Many organic compounds such as anthracene are also semiconductors (Gutmann and Lyons, 1967).

Many of the electronic properties of semiconductors can be explained with a simple model. Figure 3.2a shows the arrangement of silicon atoms in the silicon crystal. Silicon is a valence IV element and each atom has four electrons in the outermost shell. In a silicon crystal every atom has four nearest neighbor atoms, with each of which it shares two covalent electrons, thus completing the stable eight-electron shell. It takes 1.12 eV of energy, known as the *bandgap energy* of silicon, to separate an electron from the nuclei and create a free *conduction electron.* Conduction electrons are usually referred to simply as *electrons.* They can move freely and carry electric currents. The removal of an electron from a nucleus leaves behind a *hole* or a void. Electrons from neighboring atoms can fill this hole and thus cause the hole to move to a new site. This motion of the electrons naturally also carries electric current. It is easier to think of this conduction of current as due to the motion of positively charged holes in the opposite direction. The concept of a hole is analogous to that of a bubble in a liquid; although it is actually the

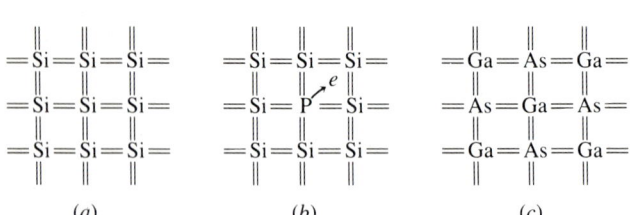

Figure 3.2 A simple model of semiconductors: (*a*) The electron bond diagram of pure Si. (*b*) The diagram of an *n*-type Si doped with phosphorous atoms. (*c*) The structure of GaAs.

liquid that moves, it is much easier to think of the motion of the bubble in the opposite direction.

In a semiconductor having bandgap energy E_g, the (conduction) electron density n and the hole density p have the following relationship:

$$np = Be^{-E_g/kT} \equiv n_i^2 \qquad (3.2.1)$$

where B is more or less a constant for all semiconductors ($B \approx 10^{39}$ cm^{-6}), k is the Boltzmann constant, and T is temperature in degrees Kelvin. In pure, or *intrinsic* semiconductors, n and p are equal and

$$n = p = n_i \qquad (3.2.2)$$

Therefore n_i is known as the *intrinsic carrier concentration*. Table 10.2 shows the E_g of some selected semiconductors.

For Si, with $E_g = 1.12$ eV, n_i is only about 10^{10} cm^{-3} at room temperature. When one considers that metals have $n > 10^{22}$ cm^{-3} it is clear that pure silicon is relatively nonconductive. The conductivity of Si may be increased by adding a small amount of group V elements such as P or As, as shown in Fig. 3.2b. After contributing four electrons to fulfill the covalent bonds, P has a fifth electron that can then easily escape from the P nucleus and become a free electron. Such impurity atoms are called *donors*, since they donate electrons. If the donor density N_d is much larger than n_i,

$$n = N_d \qquad (3.2.3)$$

and

$$p = n_i^2/n = n_i^2/N_d \qquad (3.2.4)$$

For example, a Si sample doped with 10^{15} cm^{-3} of P would have $n = 10^{15}$ cm^{-3} and $p = 10^5$ cm^{-3}. Notice that $n \gg p$ in donor-doped semiconductors, which are called *n*-type semiconductors. Here electrons are the *majority* carriers and holes are the *minority* carriers.

Similarly, if group III elements such as B or Al are added into Si, holes are created. Such dopant atoms are called *acceptors* since they accept additional electrons; and semiconductors doped with acceptors are called *p*-type semiconductors. If N_a is the acceptor density,

$$p = N_a \qquad (3.2.5)$$

and

$$n = n_i^2/N_a \qquad (3.2.6)$$

In *p*-type semiconductors, $p \gg n$. Holes are the majority carriers and electrons are the minority carriers.

The above discussion also applies to Ge, another group IV semiconductor. For another example, Fig. 3.2c shows the atomic arrangement of gallium arsenide. Ga is a group III element and As is a group V element. The most common donors of GaAs are group VI elements such as S, which substitutes for As. The most common acceptors of GaAs are group II elements such as Te, which substitutes for Ga. Si, a group IV element, can be either a donor or an acceptor in GaAs, depending on whether it substitutes for Ga or As, which in turn depends on the temperature at which Si is introduced into GaAs.

The electrical conductivity σ of a semiconductor is

$$\sigma = q\mu_n n + q\mu_p p \qquad (3.2.7)$$

where μ_n and μ_p are the electron and hole mobilities, which are two characteristic parameters of the semiconductor and, for a given semiconductor, are somewhat dependent on doping density and temperature. The conductivity is roughly proportional to the majority carrier density. Figure 3.3 shows the conductivities of p- and n-type Si. It is common to specify the dopant type and conductivity of a sample. From Fig. 3.3, the dopant density can be found.

One often encounters a graphical representation of semiconductors known as the *energy-band diagram*. The upper line in Fig. 3.4a is the conduction-band edge, representing the lowest energy that conduction electrons may have. The lower line is the valence-band edge or the lowest energy that holes can have. The two lines are, of course, separated by the bandgap energy E_g. Most electrons have energies within a few kT above the conduction-band edge. Most holes have energies within a few kT below the valence-band edge. Seeking the lowest energy point, an electron moves downward and a hole moves upward in the energy-band diagram—a fact that happens to fit the bubble analogy of holes well.

Since the conduction-band edge represents the energy of an electron at rest (zero kinetic energy), it must reflect the variation of electric potential. In Fig. 3.4b, the left end of the sample is at a potential (defined by convention for a positive charge) 2 volts higher than the right end. The *electron* potential energy then is 2 eV *lower* at the left end. The consistency of the pictures presented so far may be verified by one observation. The voltage source in Fig. 3.4b creates a field pointing to the right in the sample. Electrons would be accelerated to the left and holes to the right. This means that electrons move downhill and holes move uphill in the accompanying energy diagram, as expected.

3.3 LIGHT ABSORPTION AND CARRIER GENERATION

Each semiconductor is characterized by a bandgap energy E_g, the energy needed to free an electron from the nucleus and thus create one electron-hole pair. This energy can be supplied in the form of light. Each light photon has energy $h\nu$, where h is the Planck's constant and ν is the frequency of the light

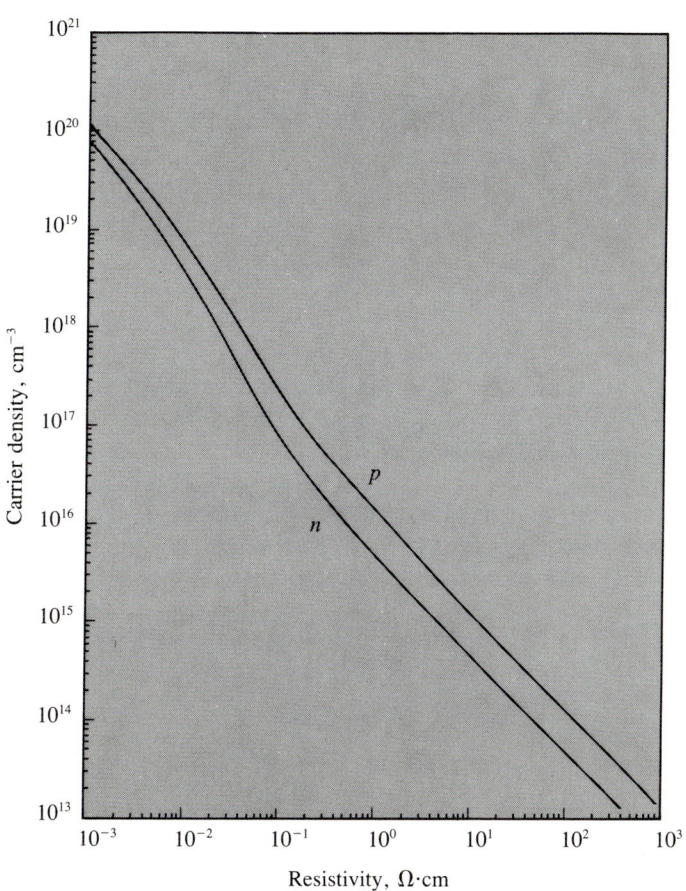

Figure 3.3 Room temperature conductivities of n- and p-type Si as functions of doping density.

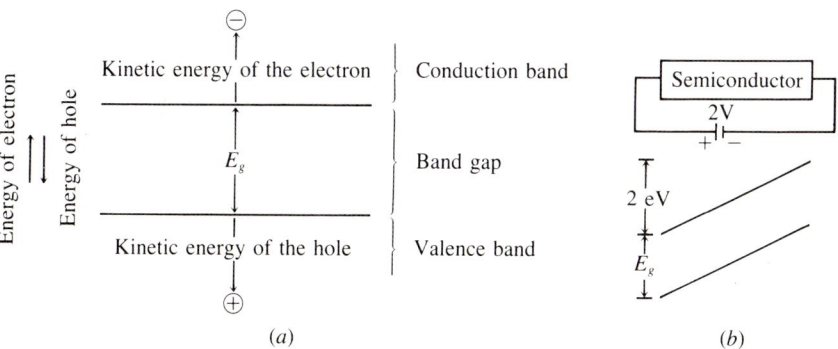

Figure 3.4 (*a*) Important concepts associated with the energy-band diagram. (*b*) The energy-band diagram reflects the electric potential variations in the sample.

44 PRINCIPLES OF SOLAR CELL OPERATION

wave. The energy $h\nu$ can be expressed as

$$h\nu(\text{eV}) = h\frac{c}{\lambda} = \frac{1.24}{\lambda(\mu m)} \qquad (3.3.1)$$

where c is the speed of light and λ is the light wavelength. For example, each light photon of 0.62 μm wavelength has 2 eV energy. Semiconductors can absorb photons having $h\nu > E_g$. The energy of each absorbed photon is consumed by raising a valence-band electron into the conduction band and creating one electron and one hole.

Each photon having $h\nu$ greater than E_g (up to several E_g) is capable of generating one and only one electron-hole pair. The excess energy is dissipated as heat. Photons with $h\nu < E_g$ are basically not absorbed by the semiconductor and cannot generate electron-hole pairs. This fact plus the spectral distribution of solar radiation lead to the conclusion that the maximum photon-absorption rate or carrier-generation rate per unit area is a function of E_g of the semiconductor, as shown in Fig. 3.5. Figure 3.5 is also a plot of the cell-current density assuming that all the light-generated carriers contribute to the cell current. The actual carrier-generation rate may be less

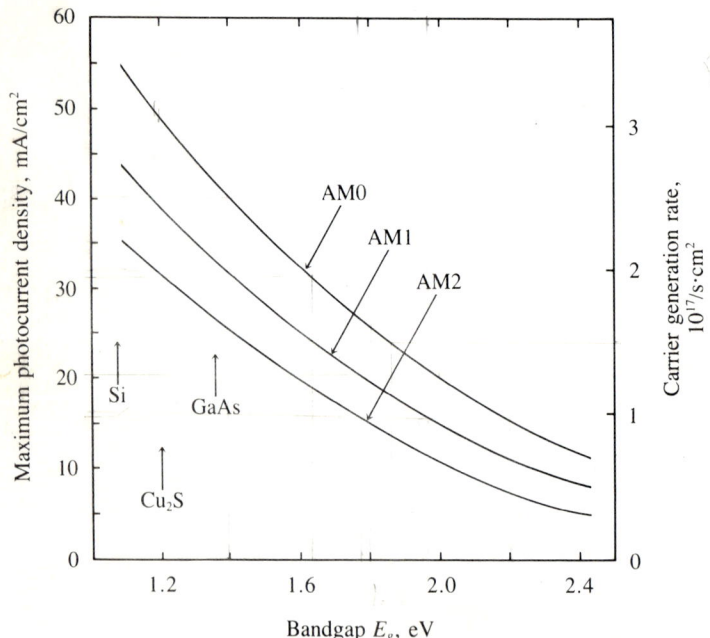

Figure 3.5 The carrier-generation rate and the maximum available cell current per unit area of cell as a function of the bandgap energy of the cell material. (See Chap. 2 for explanation of the air mass number.)

because of, for example, light reflection at the cell surface; and the actual cell current may be yet less because of incomplete collection of the carriers, as discussed in Sec. 3.6. The important fact is that cells made of narrow bandgap (small E_g) materials can be expected to generate more current.

There is a more subtle aspect to light absorption. Photons are not absorbed immediately upon incidence on the semiconductor surface. Rather, some photons can travel a considerable distance in the semiconductor before being absorbed. In general the photon flux, number of photons per square centimeter per second, decreases exponentially with the distance of travel x

$$F(x) = F(0)e^{-\alpha x} \qquad (3.3.2)$$

Therefore, the rate of photon absorption and hence the rate of carrier generation per unit volume is

$$G(x) = \frac{-dF(x)}{dx} = \alpha F(0)e^{-\alpha x} \qquad (3.3.3)$$

α is called the *absorption coefficient* and is the inverse of the average photon penetration depth. As shown in Fig. 3.6, α is a function of λ or $h\nu$. α is

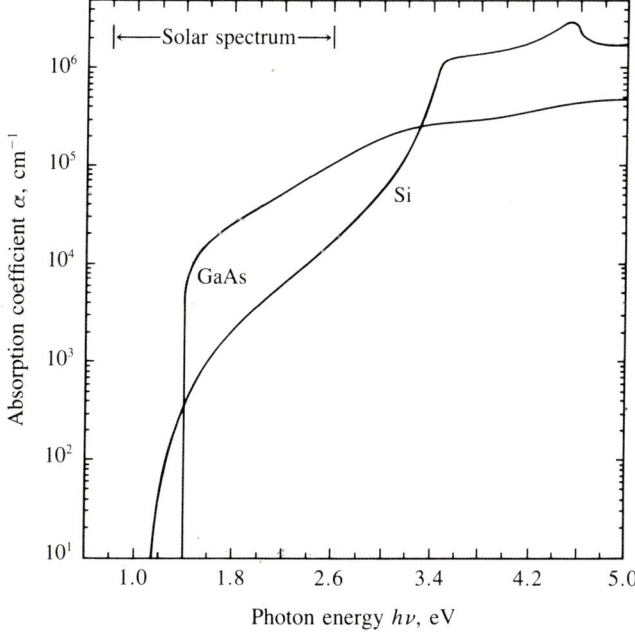

Figure 3.6 Light absorption coefficient as a function of photon energy. Si and GaAs typify the indirect- and direct-gap semiconductors.

essentially zero for $h\nu < E_g$* and larger than 10^4 cm^{-1} for large $h\nu$. In the GaAs case, this transition in α occurs rather abruptly; in Si, it occurs gradually. An abrupt change in α from 0 for $h\nu < E_g$ to a value larger than 10^4 cm^{-1} for $h\nu > E_g$ is characteristic of a class of semiconductors called the *direct-gap* semiconductors. Besides GaAs, this group includes InP, CdS, and Cu_2S. The behavior of Si is characteristic of the other group called the *indirect-gap* semiconductors, including Ge and GaP. It is difficult to predict whether a semiconductor is of the direct-gap type or the indirect-gap type since the difference is caused by the different electronic structures of the semiconductor crystals in a complicated way.

The effects are clear. Figure 3.7 shows that Si, an indirect-gap semiconductor, cannot absorb nearly all the photons with $h\nu > E_g$ until the sample is at least 100 μm thick. By contrast, GaAs, a direct-gap semiconductor, can do so with a sample 1 μm thick. Direct-gap semiconductors, then, are more suitable materials for "thin-film" solar cells. In general, less (thinner) direct-gap semiconductor material is required for making solar cells than is indirect-gap semiconductor material. This may be an important consideration in the selection of solar cell materials.

3.4 CARRIER RECOMBINATION

Echoes in a canyon eventually die out; water stirred by a cast stone will regain its calm. Nature provides means for things to return to their equilibrium states. If the carrier concentrations are made to exceed their *equilibrium* values given in Eqs. (3.2.3) to (3.2.6), such as by photo-carrier generation, the excess carriers die away by *recombination*. The rate at which *excess* electrons and holes recombine and annihilate each other in pairs is

$$\text{Recombination rate, per volume per second} = \frac{n - n_0}{\tau} = \frac{p - p_0}{\tau} (\text{cm}^{-3}\text{s}^{-1})$$

(3.4.1)

n_0 and p_0 are the equilibrium electron and hole densities; τ is the *recombination lifetime* or simply the *lifetime* of the carriers.

Long lifetimes are desirable if not essential for achieving high solar cell efficiencies. In direct-gap semiconductors of moderately high doping densities, the dominant recombination mechanism is usually the band-to-band recombination illustrated in Fig. 3.8a. Clearly, the higher the majority carrier concentration, the higher the probability of electrons' meeting and recombining with holes; thus one finds

---

*Measuring the edge of the absorption spectrum is in fact the most common way of finding the E_g of a semiconductor.

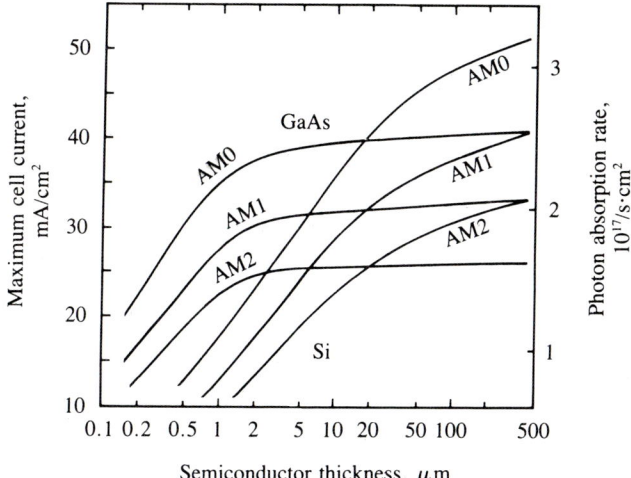

Figure 3.7 Maximum available cell current density and rate of photon absorption as function of cell thickness.

$$\tau_{\text{band-band}} = \frac{C}{\text{doping density}} \tag{3.4.2}$$

This recombination mechanism is the reverse of photon absorption. It is therefore more likely in direct-gap semiconductors, for which $C \approx 10^{10}$ s·cm^{-3}, than in indirect-gap semiconductors, for which $C \approx 10^{15}$ s·cm^{-3} (Varshni, 1967).

Figure 3.8b illustrates the Auger recombination. Since two majority carriers are involved,

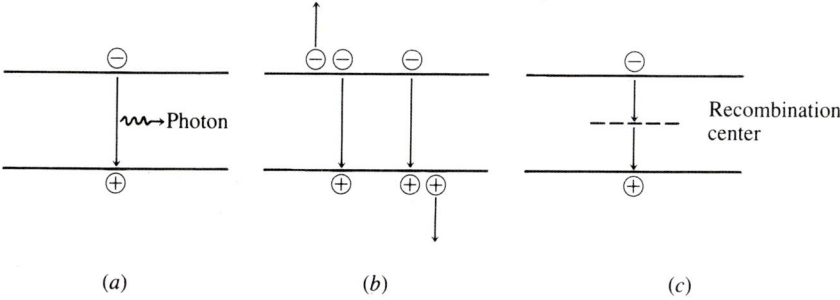

Figure 3.8 Important recombination mechanisms: (a) Band-to-band recombination is the reverse of photon absorption and important in direct-gap semiconductors. (b) Auger recombination is dominant in heavily doped semiconductors. (c) Recombination through traps is dominant in lightly doped Si.

48 PRINCIPLES OF SOLAR CELL OPERATION

$$\tau_{\text{Auger}} = \frac{D}{(\text{doping density})^2} \tag{3.4.3}$$

Auger recombination may be dominant in most heavily doped semiconductors.

For lightly doped semiconductors, recombination through recombination *centers* (also known as *deep levels* or *traps*) may dominate. This is illustrated in Fig. 3.8c. In such materials,

$$\tau_{\text{trap}} = \frac{1}{v \sigma_T N_T} \tag{3.4.4}$$

where v is the thermal velocity of the carriers, σ_T is the capture cross-section of the traps, and N_T is the trap concentration. Some chemical impurities in Si such as gold and platinum have large σ_T and therefore must be avoided in solar cells. Many semiconductor processing steps can create crystal imperfections (see Chap. 4), which behave as traps. It is possible for a silicon sample having a lifetime as long as several hundred microseconds at the beginning of the solar cell fabrication process to emerge at the end of the process with a lifetime shorter than one microsecond.

Finally, since the total recombination rate is the sum of the three separate recombination rates,

$$\frac{1}{\tau} = \frac{1}{\tau_{\text{band-band}}} + \frac{1}{\tau_{\text{Auger}}} + \frac{1}{\tau_{\text{trap}}} \tag{3.4.5}$$

Figure 3.9 shows that recombination lifetime can vary randomly over a large range due to unintentionally introduced recombination centers. The only predictable trend is that carrier lifetimes tend to fall off at high doping densities, presumably due to Auger recombination and, in the case of direct-gap semiconductors, band-to-band recombination.

Another concept closely related to recombination lifetime is diffusion length. Suppose there is a steady source of carrier generation at one plane in a sample; the minority carriers will diffuse away from the plane while recombination diminishes their numbers. The result is an excess-carrier-density profile that decreases exponentially with distance from the plane with a decay length called the *diffusion length*, L (Muller and Kamins, 1977),

$$L = \sqrt{kT\mu\tau/q} = \sqrt{D\tau}, \tag{3.4.6}$$

where $D = kT\mu/q$ is the *diffusion coefficient*. Usually L and D are accompanied with subscripts n or p to denote whether electrons or holes are being considered.

3.5 pn JUNCTIONS

All practical photovoltaic cells have intentionally built-in asymmetry, i.e., one side of the cell is different from the other side. Without some asymmetry,

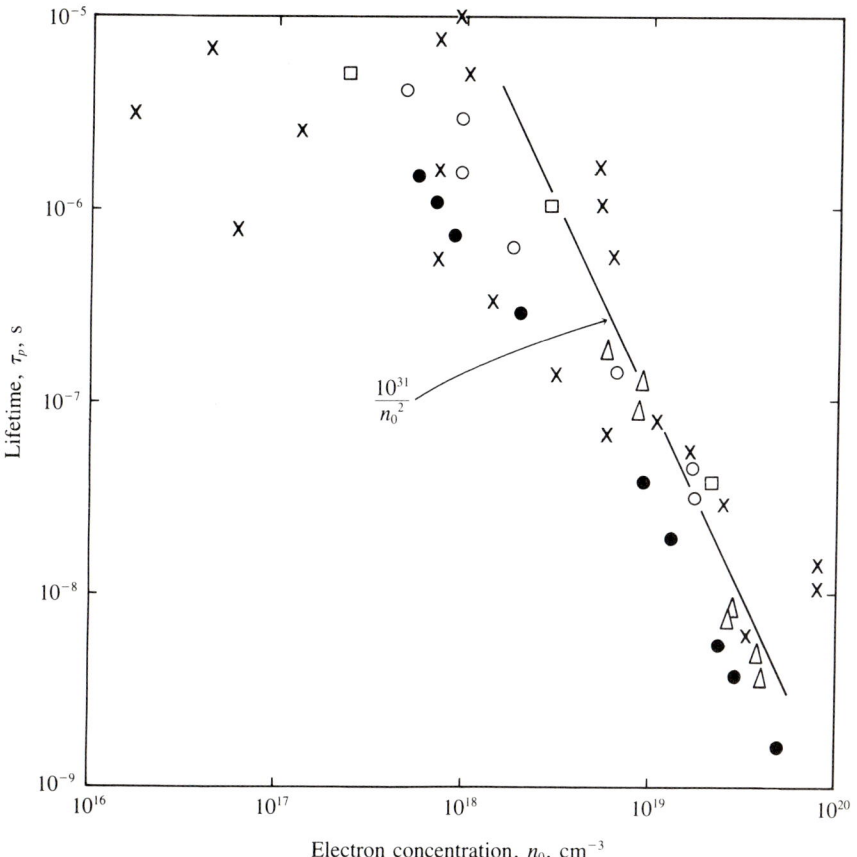

Figure 3.9 Recombination lifetime in n-type Si measured by different investigators (indicated by different symbols). At high doping densities, the lifetimes fit Eq. (3.4.3) with $D = 10^{31}\text{s}\cdot\text{cm}^{-6}$. τ_n has similar dependence on p_0.

there would be no reason for current to flow in one direction rather than the other (assuming uniform illumination in both directions) and the only reconciliation would be that no current flows in either direction. When the asymmetry consists of a p-type semiconductor on one side and an n-type semiconductor on the other, as shown in Fig. 3.10a, the structure is known as a *pn junction*. In this section, we shall discuss unilluminated *pn* junctions and remark on the general properties of some other junction structures of interest.

Due to the differences in electron and hole concentrations on the two sides, there is a strong tendency for electrons to diffuse from the n side to the p side and for holes to diffuse in the opposite direction. Such diffusions would cause positive charges to appear on the n side and negative charges to appear on the p side (Fig. 3.10a). The resultant double layer of charges sets up an electric field. The field in turn creates an internal potential drop at the junc-

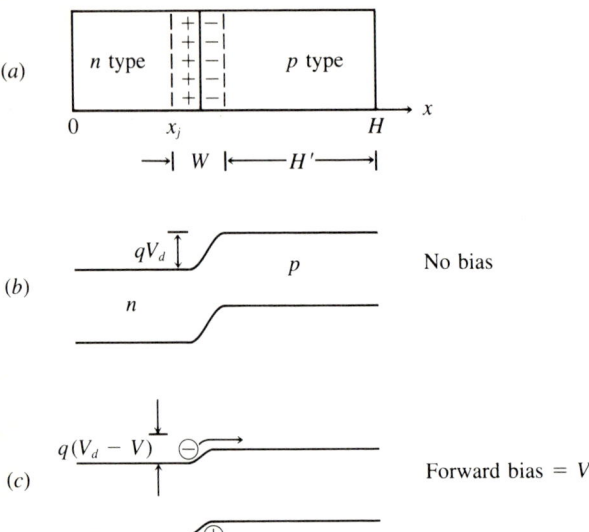

Figure 3.10 (*a*) *pn* junction. (*b*) and (*c*) Energy-band diagrams with no voltage and with V applied across the junction.

tion. This condition is illustrated in Fig. 3.10*b* by the use of an energy diagram. The tendency of the electrons to diffuse to the right is balanced by their tendency to move downward in an energy diagram (see Sec. 3.2) or to be stopped by a potential barrier. To exactly counterbalance the diffusion tendency, the built-in voltage can be shown to be (Muller and Kamins, 1977)

$$V_d = \frac{kT}{q} \ln \frac{N_d N_a}{n_i^2} \qquad (3.5.1)$$

The space containing the two charge layers is called the *depletion region* or the *space-charge region*.

When a bias voltage is applied to the junction with the *p* side positive with respect to the *n* side, the *pn* junction is *forward-biased* and the energy diagram is as shown in Fig. 3.10*c*. The potential barrier is not high enough to balance the diffusion tendency and electrons are *injected* into the *p* side and holes *injected* into the *n* side with a resultant current flow. For reference and without proof or discussion, we now list the basic equations (Muller and Kamins, 1977, or Grove, 1967, or any textbook on semiconductor electronics) that govern the behaviors of the minority carriers. (See Fig. 3.10*a* for the geometry of the junction.)

$$\frac{dJ_n}{dx} - \frac{n_p - n_{p0}}{q\tau_n} = 0 \qquad (3.5.2)$$

electrons on *p* side

$$J_n = q\mu_n n_p E + qD_n \frac{dn_p}{dx} \qquad (3.5.3)$$

3.5 pn JUNCTIONS

$$\left.\begin{array}{r}\dfrac{dJ_p}{dx} + \dfrac{p_n - p_{n0}}{q\tau_p} = 0 \\[2ex] J_p = q\mu_p p_n E - qD_p \dfrac{dp_n}{dx}\end{array}\right\} \text{holes on } n \text{ side} \quad\quad (3.5.4)$$
$$(3.5.5)$$

The boundary conditions are

$$p_n = p_{n0} e^{qV/kT} \qquad x = x_j \qquad (3.5.6)$$

$$n_p = n_{p0} e^{qV/kT} \qquad x = x_j + W \qquad (3.5.7)$$

$$S_p(p_n - p_{n0}) = D_p \frac{dp_n}{dx} - \mu_p p_n E_p \qquad x = 0 \qquad (3.5.8)$$

$$S_n(n_p - n_{p0}) = -D_n \frac{dn_p}{dx} - \mu_n n_p E_n \qquad x = H \qquad (3.5.9)$$

J is the current density, μ is the mobility, S is the surface recombination velocity, and $D = kT\mu/q$ is the diffusion coefficient. The subscripts n and p indicate electrons or holes except for n_p and p_n, which are the electron concentration on the p side and the hole concentration on the n side, respectively. The subscript 0 stands for equilibrium.

If the doping levels on the two sides are uniform, then the electric fields outside the depletion region, E_n and E_p, are negligible and the solution to Eqs. (3.5.2) through (3.5.5) is

$$p_n - p_{n0} = C_1 \cosh \frac{x}{L_p} + C_2 \sinh \frac{x}{L_p} \qquad x \le x_j \qquad (3.5.10)$$

$$n_p - n_{p0} = C_3 \cosh \frac{x}{L_n} + C_4 \sinh \frac{x}{L_n} \qquad x_j + W \le x \le H$$
$$(3.5.11)$$

Using the boundary conditions of Eqs. (3.5.6) through (3.5.9), C_1 through C_4 can be found and the injected current density (A/cm²) is

$$J_{inj} = J_0 (e^{qV/kT} - 1) \qquad (3.5.12)$$

$$J_0 = q \frac{D_p}{L_p} \frac{n_i^2}{N_d} \frac{\dfrac{S_p L_p}{D_p} \cosh \dfrac{x_j}{L_p} + \sinh \dfrac{x_j}{L_p}}{\dfrac{S_p L_p}{D_p} \sinh \dfrac{x_j}{L_p} + \cosh \dfrac{x_j}{L_p}} + q \frac{D_n}{L_n} \frac{n_i^2}{N_a} \frac{\dfrac{S_n L_n}{D_n} \cosh \dfrac{H'}{L_n} + \sinh \dfrac{H'}{L_n}}{\dfrac{S_n L_n}{D_n} \sinh \dfrac{H'}{L_n} + \cosh \dfrac{H'}{L_n}}$$
$$(3.5.13)$$

J_0 is known as the *reverse saturation current* of the junction. We shall see later that it is desirable to minimize J_0 of a solar cell. It is only necessary to remember from Eq. (3.5.13) that the first term represents hole injection into the n region and that, to minimize it, N_d and L_p should be made large and S_p small (few recombinations at the front surface). The second term represents electron injection into the p region and can be reduced by making N_a and L_n large and S_n small. Even for arbitrary and nonuniform N_a, N_d, E, τ, and D, Eq. (3.5.12) is still valid; however, it is impossible to express J_0 in a closed form in the general case.

While the current represented by Eq. (3.5.12) is the dominant one in high-quality Si pn junctions, it may not be dominant in low-quality junctions of Si or junctions of semiconductors with high E_g (therefore small n_i^2) such as GaAs. Often, another current called the *depletion-region-recombination* current is significant. This current results from the recombinations that occur within the depletion region. It can be shown (Grove, 1967) that this current

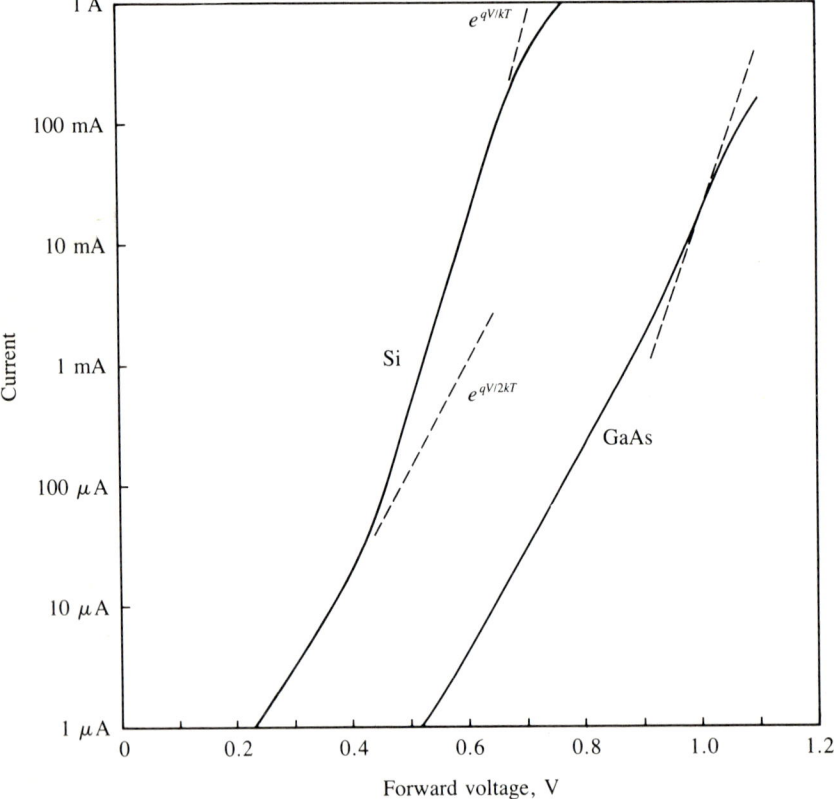

Figure 3.11 *I-V* characteristics of high-quality pn junctions at 25°C. Large area pn junctions typically show higher currents at the low voltage range.

density is

$$J_r \approx \frac{qn_i}{\tau_0} W e^{qV/2kT} \quad (3.5.14)$$

$$W = \sqrt{\frac{2\epsilon_s}{q}(V_d - V)\frac{N_a + N_d}{N_a N_d}} \quad (3.5.15)$$

ϵ_s is the permittivity of the semiconductor. Also, at high current levels, the so called *high-level-injection effect* invalidates Eq. (3.5.12) and makes J_{inj} follow a $e^{qV/2kT}$ relationship. Figure 3.11 shows some actual current-voltage curves and highlights the effects of temperature and bandgap energy. Often, over a current range the *I-V* characteristics may be fitted to

$$J = J_0(e^{qV/\gamma kT} - 1) \quad (3.5.16)$$

where γ is between 1 and 2 and J_0 and γ are determined by curve fitting. It will be seen later that good solar cell performance demands small J_0 and γ.

Figure 3.12 plots Eq. (3.5.16) on a linear current scale. Clearly, a *pn* junction passes large current in one direction only. Therefore, a *pn* junction is a *rectifying* junction. As a practical device, a *pn* junction is called a *rectifier* or a *diode*.

Many other types of junctions also exhibit the behavior represented by Eq. (3.5.16). They include certain metal/semiconductor or Schottky junctions, electrolyte/semiconductor junctions, and junctions between two dissimilar semiconductors such as Cu_2S and CdS. These junctions are interesting and attractive alternatives to *pn* junctions as solar cell structures. Any discussion of solar cells based on Eq. (3.5.16), then, applies to all types of solar cells and not just to *pn* junction solar cells.

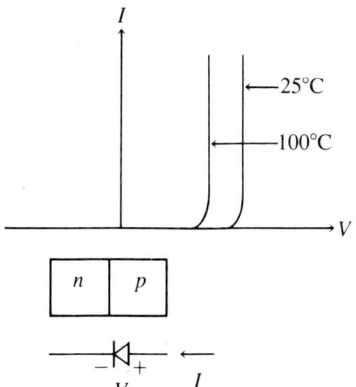

Figure 3.12 *I-V* characteristics and symbol of a rectifying diode.

3.6 SHORT-CIRCUIT CURRENT

In this section we analyze the diode current in the presence of light and introduce the important concept of the short-circuit current.

If the two terminals of a diode are connected through a conductor (short circuit), no current will flow in the circuit according to Eq. (3.5.16), since the diode voltage is zero. A delicate balance exists between the tendency of carrier flow by diffusion and the impeding effect of the potential barrier at the junction. The balance is tipped when light shines on a diode, as shown in Fig. 3.13. A current, aptly called the *short-circuit current*, I_{sc}, flows in the circuit from the p-side terminal to the n-side terminal.

The most convenient way of analyzing the short-circuit current is to divide the solar spectrum into many segments, each having a narrow range of wavelength, and find the current due to each spectral segment. Since each spectral segment is essentially monochromatic, it may be assigned a single wavelength λ and an absorption coefficient $\alpha(\lambda)$. If the number of incident photons in this spectral segment per square centimeter per second is $F(\lambda)$, the carrier-generation rate (cm^{-3} s^{-1}) [Eq. (3.3.3)] is

$$G(x) = \alpha F (1 - R) e^{-\alpha x} \qquad (3.6.1)$$

where R is the reflectivity of the solar cell surface and is in general a function of λ. The minority carrier continuity equations are

$$\frac{dJ_n}{dx} - \frac{n_p - n_{p0}}{q\tau_n} + G(x) = 0 \qquad \text{for electrons in } p\text{-type material}$$

$$(3.6.2)$$

and

$$\frac{dJ_p}{dx} + \frac{p_n - p_{n0}}{q\tau_p} - G(x) = 0 \qquad \text{for holes in } n\text{-type material}$$

$$(3.6.3)$$

J_n and J_p are given in Eqs. (3.5.3) and (3.5.5). The boundary conditions,

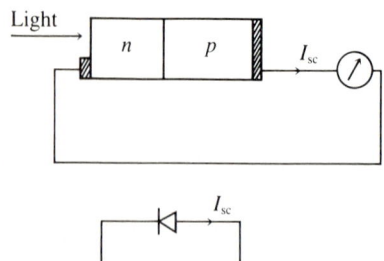

Figure 3.13 Schematic showing the direction of short-circuit current generated by light. The front layer (n-type in this illustration) is usually a diffused region; the p-region is the bulk or the base.

Eqs. (3.5.6) through (3.6.9), are applicable except that $V = 0$ (short circuit) in Eqs. (3.5.6) and (3.5.7). Again assuming zero electric fields outside the depletion region, the general solution to Eq. (3.6.3) is

$$p_n(x) - p_{n0} = A \cosh \frac{x}{L_p} + B \sinh \frac{x}{L_p} - \frac{\alpha F(1-R)\tau_p}{\alpha^2 L_p^2 - 1} e^{-\alpha x} \quad (3.6.4)$$

A and B can be determined from the boundary conditions and the resulting hole photocurrent at the junction, $x = x_j$, is

$$J_p = \left[\frac{qF(1-R)\alpha L_p}{\alpha^2 L_p^2 - 1} \right]$$

$$\frac{\dfrac{S_p L_p}{D_p} + \alpha L_p - e^{-\alpha x_j}\left(\dfrac{S_p L_p}{D_p}\cosh\dfrac{x_j}{L_p} + \sinh\dfrac{x_j}{L_p}\right)}{\dfrac{S_p L_p}{D_p}\sinh\dfrac{x_j}{L_p} + \cosh\dfrac{x_j}{L_p}} - \alpha L_p e^{-\alpha x_j}$$

(3.6.5)

This is the photocurrent due to the holes generated in the top layer of a n/p (n on p) junction solar cell and collected by the junction. Similarly, Eq. (3.6.2) can be solved to yield the photocurrent due to electrons collected at the junction from the p layer behind the junction,

$$J_n = \frac{qF(1-R)\alpha L_n}{\alpha^2 L_n^2 - 1} e^{-\alpha(x_j + W)}$$

$$\times \alpha L_n - \frac{\dfrac{S_n L_n}{D_n}\left(\cosh\dfrac{H'}{L_n} - e^{-\alpha H'}\right) + \sinh\dfrac{H'}{L_n} + \alpha L_n e^{-\alpha H'}}{\dfrac{S_n L_n}{D_n}\sinh\dfrac{H'}{L_n} + \cosh\dfrac{H'}{L_n}}$$

(3.6.6)

where x_j, W, H', and H are defined as in Fig. 3.10.

The electric field in the depletion region is so high that nearly all the light-generated carriers are swept out of the depletion region before they can recombine. The photocurrent due to the depletion region is therefore equal to the number of absorbed photons multiplied by q:

$$J_{dr} = qF(1-R)e^{-\alpha x_j}(1 - e^{-\alpha W}) \quad (3.6.7)$$

$$J_{sc}(\lambda) = J_n + J_p + J_{dr} \quad (3.6.8)$$

W is given in Eq. (3.5.15). The total short-circuit photocurrent J_{sc} is the sum of Eqs. (3.6.5), (3.6.6), and (3.6.7). These three equations can be rewritten for p/n cells by interchanging the subscript n with subscript p. The short-circuit current is a function of the cell design and is proportional to the photon

56 PRINCIPLES OF SOLAR CELL OPERATION

flux. To isolate the effects of the internal cell design parameters, one often studies the spectral response of the cell

$$\text{SR} \equiv \frac{J_{sc}(\lambda)}{qF(1-R)} \quad (3.6.9)$$

This internal spectral response is simply the ratio of the number of collected carriers to the number of photons entering the cell. When the surface reflectivity is considered a cell design parameter, one also speaks of an *external* spectral response

$$\text{SR}_{ext} \equiv \frac{J_{sc}(\lambda)}{qF} \quad (3.6.10)$$

Clearly SR is always greater than SR_{ext} and both are less than unity. The spectral response is also known as the *collection efficiency*, $\eta_{coll}(\lambda)$.

Figure 3.14 shows the calculated internal spectral response of several silicon solar cells. For wavelengths shorter than 4000 Å, photons are almost completely absorbed in the front n region. In this wavelength range, $\eta_{coll}(\lambda)$ is determined by x_j, s_p, and τ_p, which is assumed to be 3 ns. Between 5000

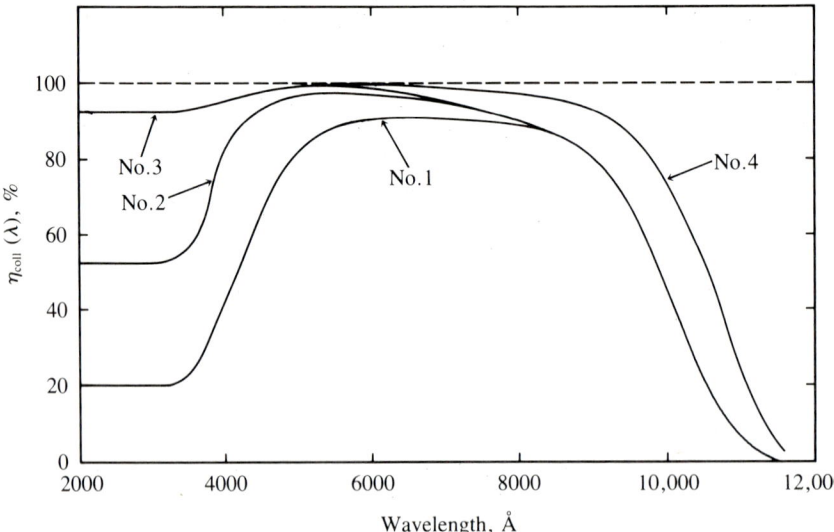

Figure 3.14 Calculated internal spectral responses of Si solar cells. Assumptions: $\tau_p = 3$ ns, $S_n \to \infty$. (Wolf, 1971.)

Å and 9000 Å, the bulk lifetime, τ_n and x_j are important. Most of the usable solar insolation falls in this spectral range. Beyond 9000 Å, photons have long penetration depths in Si. Therefore the cell thickness and the recombination velocity at the back surface, S_n, assumed to be infinite, are important.

The total cell short-circuit current is the sum of the contributions from all the spectral segments

$$J_{sc} = \sum_i qF(\lambda_i)[1 - R(\lambda_i)]SR(\lambda_i) \qquad (3.6.11)$$

The total collection efficiency is

$$\eta_{ext} = J_{sc}/q \sum_i F(\lambda_i) \qquad (3.6.12)$$

$q\sum_i F(\lambda_i)$ is simply the maximum available short-circuit current (plotted in Fig. 3.6). Typical solar cells collect 60 to 90 percent of the maximum possible current.

3.7 EFFICIENCY

In Sec. 3.5 we have analyzed the diode current under voltage bias in the dark, and in Sec. 3.6, the diode current in the light without bias (short-circuit). When both light and voltage bias are present, the total diode current is the sum of I_{sc} and the dark diode current.

$$I(V) = I_{sc} + \text{dark diode current as function of } V$$
$$= I_{sc} - I_0[e^{(qV/\gamma kT)} - 1] \qquad (3.7.1)$$

following Eq. (3.5.16). Figure 3.15a illustrates this summation. If light intensity varies, I_{sc} varies and the I-V curve would move as a whole up and down in the figure. Figure 3.15b is the equivalent circuit of a solar cell. The output voltage and current of a solar cell is determined only when a load is connected to the solar cell. An ideal battery load, which has fixed terminal voltage, is assumed in Fig. 3.15c, while a resistive load is assumed in Fig. 3.15d. When the load is a short circuit ($V_{out} = 0$), I_{out} is simply I_{sc}. When the load is an open circuit ($I_{out} = 0$), the corresponding V_{out} is called the *open-circuit voltage*, V_{oc}. From Eq. (3.7.1)

$$V_{oc} = \frac{\gamma kT}{q} \ln\left(\frac{I_{sc}}{I_0} + 1\right) \qquad (3.7.2)$$

The power output is

$$P_{out} = V_{out} \times I_{out} \qquad (3.7.3)$$

By observation of the I-V curve, one expects that a particular pair of V_{out} and I_{out} called V_m and I_m will maximize P_{out}.

$$\max P_{\text{out}} \equiv P_m = V_m \times I_m \qquad (3.7.4)$$

V_m can be found by substituting I_{out} in Eq. (3.7.3) with $I_{sc} - I_0 \times [e^{(qV_{\text{out}}/\gamma kT)} - 1]$, differentiating P_{out} with respect to V_{out} and setting the result equal to zero. Once V_m is known, I_m and P_m can be calculated. Depending on $qV_{oc}/\gamma kT$, V_m is typically 75 to 90 percent of V_{oc}, and I_m is typically 85 to 97 percent of I_{sc}. It is useful to define a *fill factor*,

$$\text{FF} = \frac{P_m}{V_{oc} \times I_{sc}} \qquad (3.7.5)$$

which is just the ratio of the area of the largest rectangle that can fit under the *I-V* curve to the product $V_{sc} \times I_{oc}$. The maximum possible power output can now be expressed as

$$P_m = V_{oc} \times I_{sc} \times \text{FF} \qquad (3.7.6)$$

FF is a function of $qV_{oc}/\gamma kT$, or $\ln[(I_{sc}/I_0) + 1]$ as shown in Fig. 3.16.
The power conversion efficiency of a solar cell is

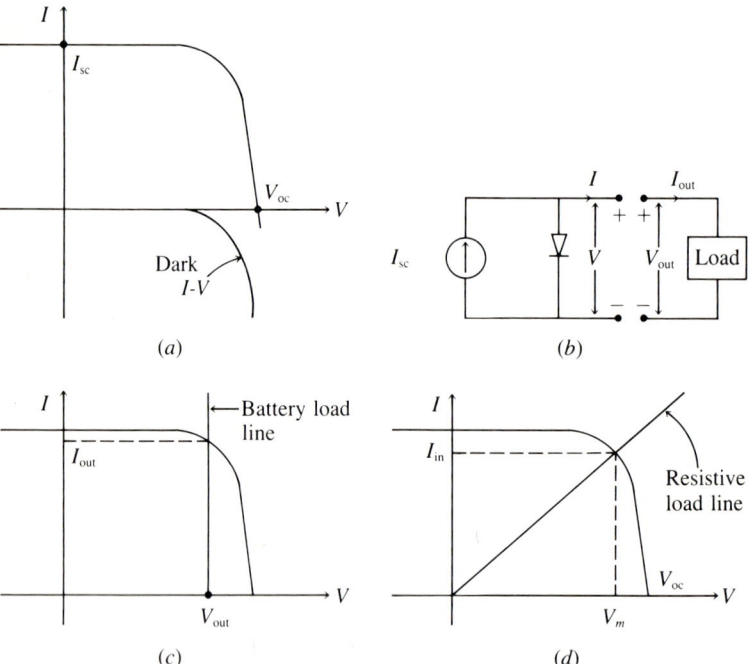

Figure 3.15 (*a*) By superposition, I_{sc} plus the dark current yields the *I-V* curve of a solar cell. (*b*) Cell *I-V* curve and load determine the output current and voltage. (*c*) An ideal battery load. (*d*) A resistance load.

LIGHT SENSORS; PHOTODIODES AND PHOTOCONDUCTORS

Referring to Fig. 3.15a, a short-circuited or reverse-biased diode (negative V) passes no (or negligible) current in the dark, and in the light passes I_{sc}, which is proportional to the light intensity. By measuring the reverse current one can gauge the light intensity. When used in this fashion the device is called a *photodiode*.

Photodiodes, often made of silicon, are used, for example, as the detectors or receivers in the emerging fiber-optical communication systems. In such systems, signals are transmitted as variations in light intensity through glass fibers as thin as human hair. The glass is of such clarity that light is attenuated by as little as 10 percent in traveling through a 1-km length of the fiber.

Another common type of light sensor is the photoconductor. Photo-carrier generation increases the electron and hole densities in a semiconductor and thus increases its conductivity (see Eq. 3.5.7). By measuring the conductance change (change of current under constant applied voltage), one can sense the light intensity. CdS photoconductor sensors are widely used in the exposure systems of cameras. Unlike Si, CdS does not respond to infrared light and thus matches human eyes more closely.

$$\eta \equiv \frac{P_m}{P_{in}} = \frac{V_{oc} \times I_{sc} \times \text{FF}}{\text{incident solar power}} \quad (3.7.7)$$

In order to achieve high conversion efficiency, it is clearly desirable to have

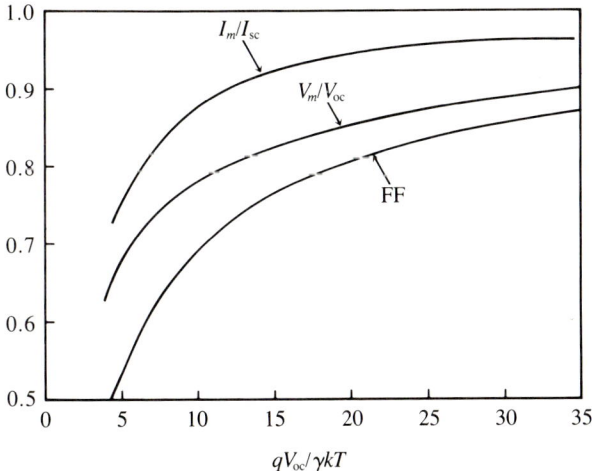

Figure 3.16 V_m/V_{oc}, I_m/I_{sc}, and FF as functions of $qV_{oc}/\gamma kT$. Note that $qV_{oc}/\gamma kT = \ln(I_{sc}/I_0 + 1)$.

60 PRINCIPLES OF SOLAR CELL OPERATION

high I_{sc} or high collection efficiency, high V_{oc} or low dark current, and high fill factor or a sharp corner in the I-V curve.

Present-day Si solar cells have efficiencies between 12 and 17 percent under one sun (AM1) illumination. Figure 3.17 is an accounting sheet of the energy losses (Wolf, 1971). In the solar spectrum, 26 percent of the energy is in the photons having $h\nu < 1.1$ eV. Of the remaining useful photons, most have energy much larger than 1.1 eV. The excess energy is wasted since there is no difference between the electron–hole pairs generated by a 1.1-eV photon and by a 2.5-eV photon. Forty-one percent of the energy in photons is so wasted. Even an ideal cell can only turn the remaining $0.78 \times 0.59 = 0.44$ or 44 percent into electricity. Depending on the cell design and the material parameters, real cells have efficiencies between 12 and 17 percent.

3.8 FACTORS AFFECTING THE CONVERSION EFFIENCY

Bandgap Energy

The relationship between V_{oc} and E_g can be illustrated by assuming that I_{inj} in Eq. (3.5.12) is the dominant diode current

$$V_{oc} \approx \frac{kT}{q} \ln \frac{J_{sc}}{J_0} \qquad (3.8.1)$$

From Eqs. (3.5.13) and (3.2.1)

$$J_0 = n_i^2 \left(\frac{C_p}{N_d} + \frac{C_n}{N_a} \right)$$

$$= B e^{-E_g/kT} \left(\frac{C_p}{N_d} + \frac{C_n}{N_a} \right) \qquad (3.8.2)$$

where the definitions of C_p and C_n are obvious from Eq. (3.5.13). Substituting (3.8.2) into (3.8.1)

$$V_{oc} \approx \frac{E_g}{q} - \frac{kT}{q} \ln \left[\frac{B}{J_{sc}} \left(\frac{C_p}{N_d} + \frac{C_n}{N_a} \right) \right] \qquad (3.8.3)$$

Since it is logarithmic, the second term in Eq. (3.8.3) is more or less a constant and is typically 0.5 V. Thus, Si has $E_g = 1.12$ eV and silicon solar cells have $V_{oc} \approx 0.55$ V; GaAs has $E_g = 1.43$ eV and GaAs solar cells have $V_{oc} \approx 0.9$ V.

V_{oc} increases with increasing E_g. On the other hand, in Fig. 3.7 we have seen that the maximum possible J_{sc} decreases with increasing E_g. As a result, we may expect solar cell efficiency to peak at a certain E_g. This is borne out by Fig. 3.18, which plots calculated conversion efficiencies for unity collection efficiency, 10^{17} cm^{-3} doping concentration and typical lifetimes.

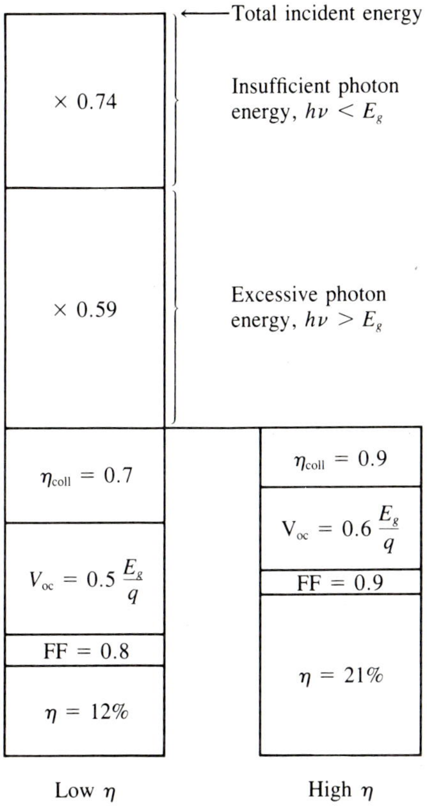

3.8 FACTORS AFFECTING THE CONVERSION EFFICIENCY 61

Figure 3.17 Accounting of the energy losses in a typical low efficiency and a hypothetical, very high efficiency, Si solar cell.

Low η column:
× 0.74 — Insufficient photon energy, $h\nu < E_g$
× 0.59 — Excessive photon energy, $h\nu > E_g$
$\eta_{coll} = 0.7$
$V_{oc} = 0.5 \frac{E_g}{q}$
FF = 0.8
η = 12%

High η column:
$\eta_{coll} = 0.9$
$V_{oc} = 0.6 \frac{E_g}{q}$
FF = 0.9
η = 21%

It is worth noting that the "maximum" efficiencies shown in Fig. 3.18 can be exceeded in concentrator systems (see Chap. 5) and in systems where several solar cell materials (see Chap. 12) or materials having graded bandgaps (see Chap. 11) are employed.

Temperature

Figure 3.18 also shows that η decreases with increasing temperature. A reading of Sec. 3.6 would show that I_{sc} is insensitive to T. It is V_{oc} that is responsible for the temperature dependence. From Eq. (3.8.3)

$$\frac{dV_{oc}}{dT} = \frac{1}{q}\frac{dE_g}{dT} - \frac{k}{q}\ln\left[\frac{B}{J_{sc}}\left(\frac{C_p}{N_d} + \frac{C_n}{N_a}\right)\right]$$

$$= \frac{1}{q}\frac{dE_g}{dT} - \frac{1}{T}\left(\frac{E_g}{q} - V_{oc}\right) \qquad (3.8.4)$$

For Si, $dE_g/dT = -0.0003$ eV/°C, $E_g/q - V_{oc} \approx 0.5$ V, therefore

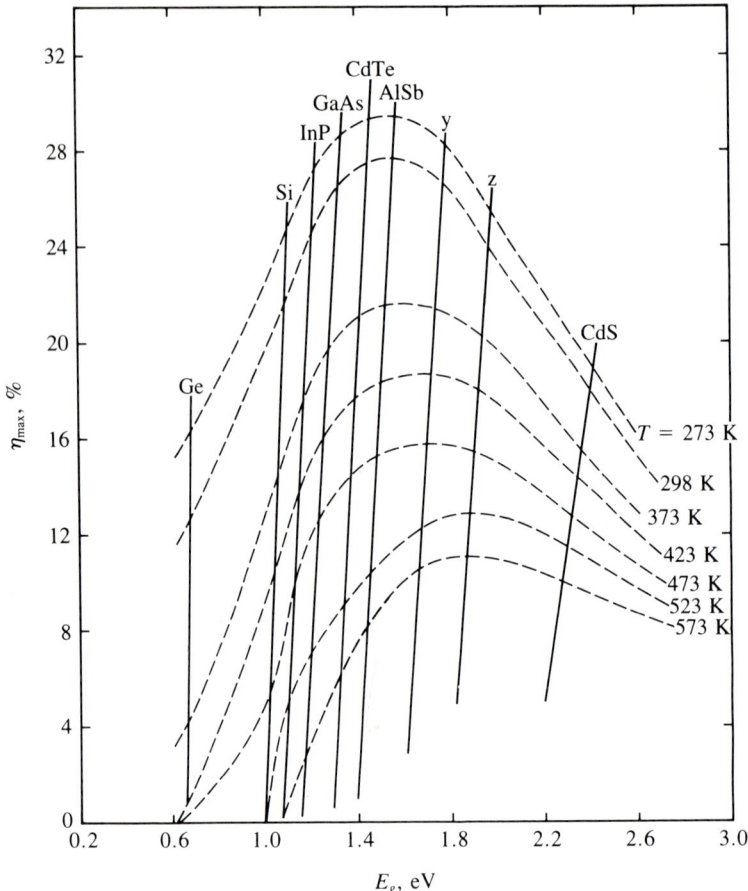

Figure 3.18 Calculated upper limits of efficiency. Assumptions: Unity collection efficiency, 10^{17} cm^{-3} doping, typical lifetimes. These limits may be exceeded by using light concentration or employing more than one solar cell material. *(Rappaport, 1959.)*

$dV_{oc}/dT = -2$ mV/°C. For every 1°C increase in temperature, V_{oc} drops by about $0.002/0.55 \approx 0.4\%$ of its room temperature value, and hence η drops by about the same percentage. For example, a Si cell having 20% efficiency at 20°C will only be about 12% efficient at 120°C. For GaAs, V_{oc} drops by 1.7 mV or 0.2% per °C. These calculated values agree well with measured temperature sensitivities. In Eq. (3.8.4), the dE_g/dT term can usually be neglected. dV_{oc}/dT, then, can be predicted from the knowledge of E_g and V_{oc} of any given cell.

Recombination Lifetime

Long carrier-recombination lifetimes are desirable mainly because they help achieve large I_{sc}. In indirect-gap materials such as Si, significant numbers of carriers are generated as far as 100 μm from the junction, and recombination lifetimes longer than 1 μs are desirable. In direct-gap materials such as GaAs or Cu_2S, recombination lifetimes as short as 10 ns may suffice. Long lifetimes also reduce the dark current and increase V_{oc}.

The key to achieving long recombination lifetimes is to avoid introducing recombination centers during material preparation and cell fabrication. Proper but often elusive "gettering" steps in the fabrication process can improve lifetimes by removing the recombination centers.

Light Intensity

In Chap. 5 we shall discuss the concept of focusing sunlight on solar cells so that a small solar cell can produce a large amount of electric power. Assuming that the sunlight intensity is concentrated by X times, the input power per cell area and J_{sc} both increase by X times. V_{oc}, according to Eq. (3.8.3), also increases by $(kT/q) \ln X$, e.g., 0.12 V for $X = 100$. The output power therefore increases by more than X times and the conversion efficiency is higher as a result of sunlight concentration.

Doping Density and Profile

Another factor that can have a significant effect on V_{oc} is the doping density. Although N_d and N_a appear in the logarithmic term in Eq. (3.8.3) they can also be easily changed by orders of magnitude. The higher the doping densities, the higher is V_{oc}. A phenomenon known as the *heavy doping effect* has recently received much attention. Because of the deformation of the band structure and the change in electron statistics at high doping concentrations, N_d and N_a in all the equations should be replaced by $(N_d)_{\text{eff}}$ and $(N_a)_{\text{eff}}$, which are shown in Fig. 3.19a. Since $(N_d)_{\text{eff}}$ and $(N_a)_{\text{eff}}$ exhibit peaks, it may not be advantageous to use very high N_d and N_a, particularly since the lifetimes tend to decrease at high doping densities, probably due to Auger recombination as discussed in Sec. 3.4. Figure 3.19b illustrates this point.

At the present time, the base doping density is usually about 10^{16} cm^{-3} in Si solar cells and about 10^{17} cm^{-3} in cells of direct-gap materials. The front diffused region is usually doped to higher than 10^{19} cm^{-3} in order to minimize the series resistance, which will be discussed later. The heavy doping effect is thus more important in the diffused region.

When N_a and N_d, or better, $(N_a)_{\text{eff}}$ and $(N_d)_{\text{eff}}$, are not uniform but decrease

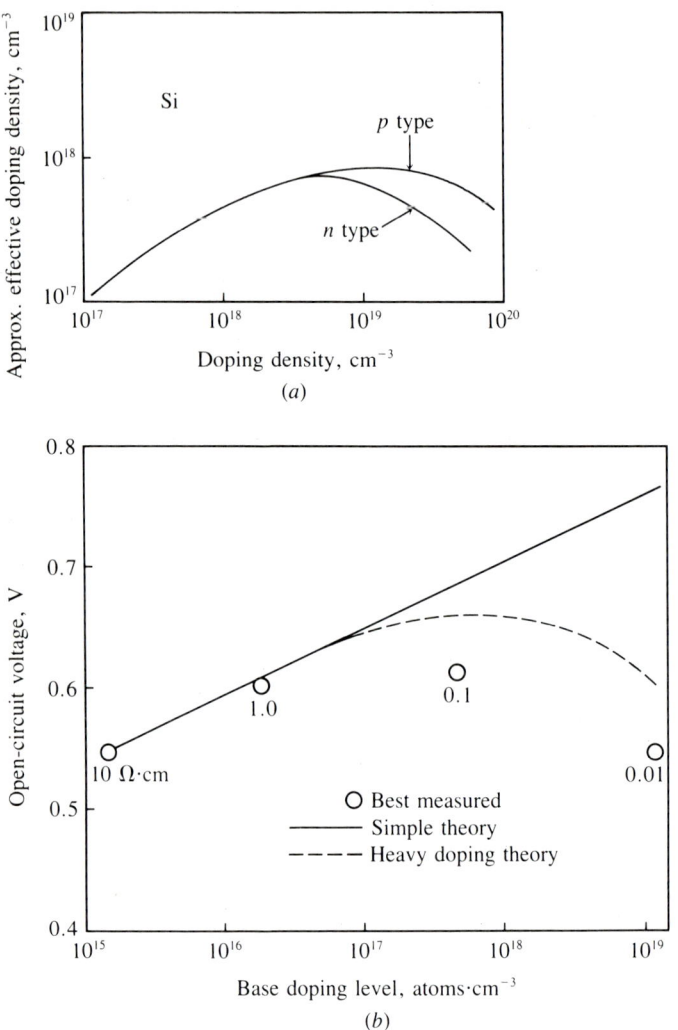

Figure 3.19 (a) The high doping effect: The effective doping density saturates or even drops with increasing doping density.
(b) V_{oc} as a function of the bulk doping density. *(Godlewski et al., 1977.)*

toward the junction, electric fields are set up in such directions as to aid the collection of photo-generated carriers. These fields improve I_{sc}. Such nonuniform doping profiles are usually impractical in the base region of the cell. They are natural in the diffused region—provided that attention is paid to the heavy doping effect.

Surface Recombination Velocities

Low surface recombination velocities help enhance I_{sc}. They also improve V_{oc} by reducing I_0. The recombination velocity at the front surface is difficult to measure and is often assumed to be infinite (a perfect trap of carriers). If the back of the cell is alloyed to a metal contact, the recombination velocity at the back surface would also be infinite. In recent years, however, a design called the *back-surface field* (BSF) cell (Mandelkorn et al., 1973) has become popular. Figure 3.20 shows the structure. An additional p^+ layer is diffused into the back side of the cell before the metal contact is deposited. As the energy diagram in Fig. 3.20 shows, the p/p^+ interface presents a barrier to the electrons that would have easily reached the ohmic contact and recombined there. It can be shown that the recombination velocity at the p/p^+ interface, which is now considered the back surface for evaluating Eqs. (3.5.13) and (3.6.6), can be expressed as (Hauser, 1975)

$$S_n = \frac{N_a}{N_{a^+}} \frac{D_{n^+}}{L_{n^+}} \coth \frac{W_{p^+}}{L_{n^+}} \qquad (3.8.5)$$

where N_{a^+}, D_{n^+}, and L_{n^+} are the doping density, diffusion coefficient, and diffusion length in the p^+ region. If $W_{p^+} = 0$, then $S_n = \infty$ as expected. However, if W_{p^+} is comparable to L_{n^+} and $N_{a^+} \gg N_a$, then S_n can be assumed to be zero. The effects of S_n on J_{sc}, V_{oc}, and η are illustrated in Fig. 3.21. Note that both J_{sc} and η show a peak when S_n is small.

Series Resistance

In any real solar cell there is some series resistance, which can arise from the lead, the metal contact grid, or the bulk cell resistance. However, the domi-

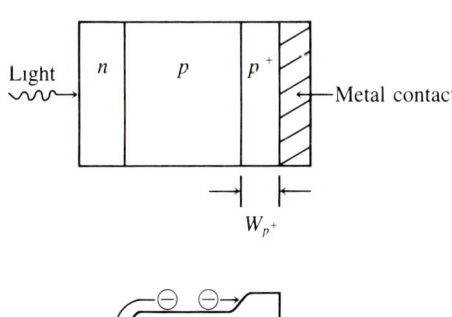

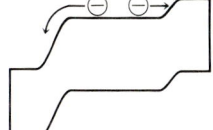

Figure 3.20 Back surface field cell. The field at the p/p^+ junction retards the flow of the electrons toward the back surface.

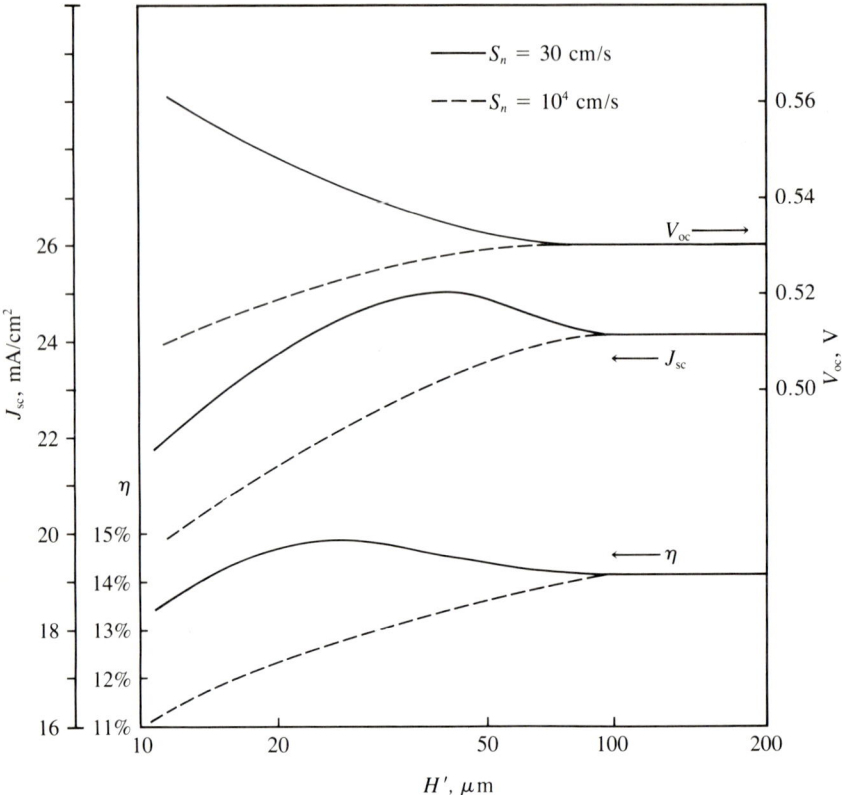

Figure 3.21 Effects of the back surface recombination velocity on cell performance. Note that η peaks at certain cell thicknesses when S_n is small.

nant source of series resistance is usually the thin diffused layer. Current collected at the *pn* junction in Fig. 3.1 must flow in the *n* layer to the closest metal line, and this is a resistive path. Clearly, the series resistance can be minimized by spacing the metal lines closely.

The effect of a finite series resistance, R_s is to shift the *I-V* curve in, say, Fig. 3.15d to the left by $I \times R_s$. Roughly speaking, this has the effect of reducing V_m by $I_m \times R_s$. If $I_m \times R_s \ll V_m$, the impact on η is minimal. This is usually the case except when the material has very high resistivity, such as with many amorphous and organic semiconductors, or when I_m is large, as in a concentrator system.

Metal Grid and Optical Reflection

The metal contact grid on the front surface is opaque to sunlight. To maximize I_{sc}, the area occupied by the metal grid should be minimized. In order to keep

R_s low, the general rule is to make the metal grid in the form of closely spaced, fine lines.

Not all the light can enter Si because of optical reflection. The reflectivity of a bare Si surface is about 40 percent. It can be reduced by the use of antireflection coatings. For a single-frequency light incident on the cell at a right angle, the reflectivity can be made zero by a single coating one-fourth wavelength thick and having an index of refraction equal to $\sqrt{n}$, where n is the index of refraction of Si. More than 90 percent of the usable photons in the solar spectrum can be transmitted into Si when a 600-Å coating of Ta_2O_5 is applied; 84 percent is transmitted with a 1000-Å coating of SiO_2. Better results can be achieved with multiple-layer coatings (Wang et al., 1973). Less is transmitted when light is incident at an oblique angle.

3.9 SUMMARY

Many structures, such as *pn* junctions and metal-semiconductor junctions, exhibit a distinctive rectifying *I-V* characteristic as shown in Fig. 3.12. Under illumination, their *I-V* curves are shifted parallel to the current axis as shown in Fig. 3.15*a*. Connected to a load, the device would operate in the quadrant of positive voltage and positive current (Fig. 3.15), thus providing power to the load.

The conversion efficiency of a solar cell is the ratio of its output electrical power to the input light power. To achieve high efficiency, it is desirable to have large short-circuit current, high open-circuit voltage, and large fill factor. The short-circuit current is higher if the solar cell is made from materials with small energy gap, E_g. Good fabrication processes and good cell design can also increase the short-circuit current by minimizing carrier recombination. Solar cells made of material having a large E_g tend to have higher open-circuit voltage. Fill factor is a measure of the sharpness of the knee in the *I-V* curve. It can be lowered by the presence of series resistance and tends to be higher whenever the open-circuit voltage is higher. The conversion efficiency increases with increasing light intensity and decreasing temperature.

The highest efficiencies are expected from solar cells made from materials with E_g between 1.2 and 1.6 eV. For thin-film cells, direct-gap semiconductors, which absorb photons near the surface, are preferable. The most common solar cells, silicon *pn*-junction cells, have efficiencies ranging between 12 and 17 percent.

REFERENCES

Godlewski, M. P., Brandhorst, H. W., Barona, C. R. (1977), Record, 12th IEEE Photovoltaic Spec. Conf., 32–36.

Grove, A. S. (1967), *Physics and Technology of Semiconductor Devices*, John Wiley & Sons.
Gutmann, F., and Lyons, L. E. (1967), *Organic Semiconductors*, John Wiley & Sons.
Hauser, J. R. and Dunbar, P. M. (1975), *Solid State Elec.*, Vol. 18, 715.
Hovel, H. J. (1975), *Semiconductors and Semimetals, Vol. 11—Solar Cells*, Academic Press.
Mandelkorn, J., Lamneck, J. H., and Scudder, L. R. (1973), Record, 10th IEEE Photovoltaic Spec. Conf., 207–211.
Muller, R. S. and Kamins, T. (1977), *Device Electronics for Integrated Circuits*, Wiley-Interscience.
Rappaport, P. (1959), *RCA Rev.*, Vol. 20, September, 373–397.
Varshni, Y. P. (1967), *Phys. Stat. Sol.*, Vol. 19, 459–514.
Wang, E. Y., Yu, F. T. S., and Sims, V. L. (1973), Record, 10th IEEE Photovoltaic Spec. Conf., 168–173.
Wolf, M. (1971), *Energy Conversion*, Vol. 11, June, 63–73.

PROBLEMS

3.1 *Solar cell current versus bandgap* What is the relationship between the left- and right-hand vertical scales of Fig. 3.5? In your own words explain why cells made of narrow bandgap materials may be expected to generate more current.

3.2 *Simiplification of diode current expression*
    (a) Find the expression for J_0 (3.5.13) in the limiting case of $S_p = S_n = \infty$.
    (b) Further simplify the expression by assuming $X_j \gg L_p$ and $H' \gg L_n$. The resultant expression should be familiar if you have studied the theory of *pn* junctions.

3.3 *Derivation of solar cell I-V expression*
    (a) Show that Eq. (3.6.4) satisfies Eq. (3.6.3).
    (b) Find A and B in Eq. (3.6.4) from the boundary conditions.
    (c) Using the results of (b), verify Eq. (3.6.5).

3.4 *Cell efficiency* In your own words, discuss Fig. 3.14.

3.5 *Calculation of cell efficiency* Assuming that a Si solar cell has $J_{sc} = 35$ mA/cm^2 under AM1, from Fig. 3.5 find η_{coll}. Assuming that its $J_0 = 10^{-11}$ mA/cm^2, find V_{oc} and FF from Fig. 3.16 and its caption. Finally, follow the example of Fig. 3.17 to find the efficiency of this cell. Note that η can be found more directly from $P_{out} = J_{sc} \times V_{oc} \times$ FF and the AM1 radiation intensity given in Chap. 2. Try it.

3.6 *Temperature effects* Derive Eq. (3.8.4). Using the information in Fig. 3.18, plot two curves of η versus T for Si and GaAs, respectively. Compare them with the discussion in the text following Eq. (3.8.4).

3.7 *Factors affecting cell efficiency* Are J_{sc}, V_{oc}, and η improved, degraded, or not significantly affected when each of the following is increased: T, τ, E_g, S, sunlight concentration ratio, N_a and N_p, cell thickness, series resistance, cell reflectance?

CHAPTER
FOUR
MATERIALS AND PROCESSING

CHAPTER OUTLINE

4.1 MATERIAL PROPERTIES AND PROCESSING TECHNIQUES
BOX: MAKING AN INTEGRATED CIRCUIT
4.2 CONVENTIONAL SILICON CELL PROCESSING
4.3 PROCESSING CADMIUM SULFIDE CELLS
4.4 ENVIRONMENTAL AND OTHER CONSIDERATIONS
4.5 SUMMARY
REFERENCES
PROBLEMS

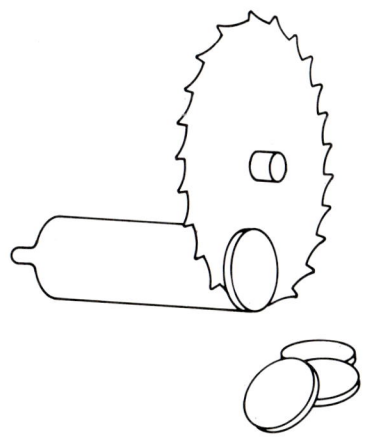

Terrestrial solar cells made conventionally from single-crystal silicon are constructed as shown in Fig. 4.1. Typical *n*-on-*p* cells have been made from round silicon wafers about 0.3 mm thick. The unilluminated bottom, or back, side of a cell has a metal coating that contacts the *p*-type body of the silicon. An *n*-type top layer which forms the *pn* junction is heavily doped for low resistivity. Metal fingers about 0.1 mm wide and 0.05 mm thick make ohmic contact to this front layer to collect the current. A transparent insulating antireflection coating approximately 0.06 microns (μm) thick covers the top silicon layer and provides much better transmission of light than can be obtained with bare silicon.

Comparing this structure with that of an integrated circuit (IC), one is struck with the relative simplicity of the solar cell. The integrated circuit contains thousands of *pn* junctions in its transistors. The major features of it are only a few microns across, and the operations it performs are quite varied and complex when compared with those of the solar cell. Silicon fabrication techniques are well understood, and the processes used to make an integrated circuit can be used to make a solar cell as well. The reader may wonder why an entire chapter should be devoted to solar cell materials and processing.

4.1 MATERIAL PROPERTIES AND PROCESSING TECHNIQUES

It is true that silicon cells have been made for terrestrial applications by employing the conventional cell design and traditional IC processing methods. These have been relatively expensive cells, however, costing more than

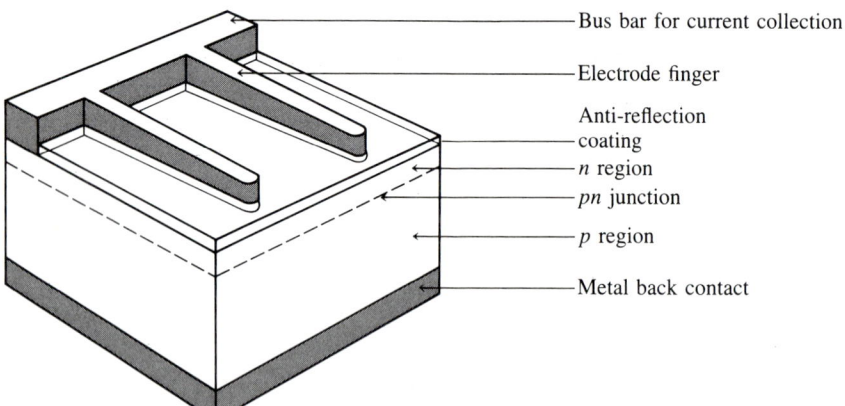

Figure 4.1 Schematic view of typical commercial single-crystal silicon solar cell.

MAKING AN INTEGRATED CIRCUIT

To create the thousands of transistors in an integrated circuit requires forming in a silicon wafer precisely dimensioned n-type and p-type regions containing different impurity atoms. The typical steps illustrated below to make a single pn junction in p-type silicon may be repeated several times during the manufacture of a complete integrated circuit.

One begins by entering in a computer file the precise locations of all the n-type regions in the circuit. The computer then controls an electron beam that draws the pattern by selectively exposing a film of electronresist (similar to photoresist described below). By chemical etching, the pattern of electronresist is transferred to an underlying thin chromium film on a glass plate. The glass plate carrying the chromium pattern (called a *photomask*) is slightly larger than the silicon water, whose diameter may range from 5 to more than 12 cm.

The surface of the silicon wafer is oxidized in a furnace heated to about 1000°C and having dry oxygen or steam flowing through it. One then makes openings in this oxide to permit dopant atoms to enter the silicon. The openings are made by first coating the oxide with a liquid photosensitive polymer, called *photoresist*, which is then hardened by baking and exposed with ultraviolet light through the photomask [sketch (a)]. Where the light reaches the photoresist, the polymer crosslinks strongly so that when the wafer is placed in a bath of developer the portions of the resist that were illuminated continue to adhere to the oxide. The resist washes away from each shaded region such as that under the dark line in the pattern, leaving openings through which the underlying oxide can be removed in a chemical bath or in a dry-plasma-etching process.

The remaining photoresist is then removed and [sketch (b)] dopant atoms enter the silicon through the opening in the oxide to change the silicon there to

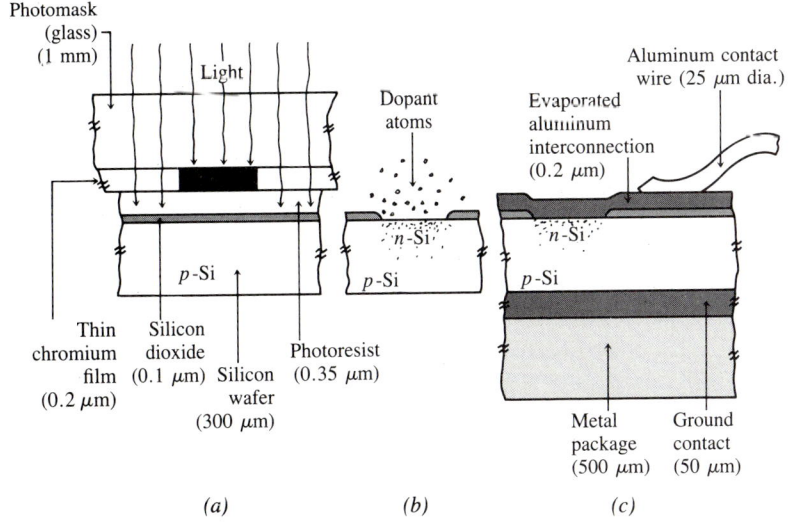

(a)   (b)   (c)

> n-type. This doping may be done in a 1000°C furnace in which a gas containing phosphorus is flowing. Alternatively one may implant phosphorus ions accelerated in a vacuum to energies from tens to hundreds of thousands of electron-volts. After the implantation, a thermal annealing process eliminates structural damage and activates the dopant atoms so they produce mobile charge carriers.
>
> These steps are repeated many times to make multiply doped and electrically connected regions that form the circuit. Additional layers of insulating oxides, and perhaps polycrystalline silicon, may be formed and patterned with other photomasks. Finally, an aluminum conductive layer is evaporated and patterned with yet another photoresist layer [sketch (*c*)]. The 0.4-cm-square circuit chips are now separated mechanically and put in individual packages. Connecting wires are attached to the accessible aluminum "pads" by ultrasonic bonding, and the circuit is ready for testing and use.

10 dollars per peak watt of output, and so have been most suitable for remote uses such as powering isolated communication equipment located where the cost of electricity from other sources is high. Two key factors affect the choices of cell materials and processing methods:

- *Cost of electrical energy produced.* The cost of output from a photovoltaic system—for example, in dollars per kWh—is determined by cell and array efficiencies, and all the costs incurred while fabricating, installing, and operating the system. To these must be added balance-of-system (BOS) costs, such as the cost of the land devoted to the system, and the cost of power conditioning and energy storage.
- *Energy payback time, or ratio.* Energy is used at each stage in the manufacture of a photovoltaic power system—in extracting the raw materials from the ground, in refining, in shaping the materials, and so on. The length of time that the system must operate to produce electrical energy equal to the total energy used in its manufacture, called the *energy payback time,* should be no more than a few years.

The *energy payback ratio,* the total energy produced during its useful life divided by the total energy used to make the system, must be greater than unity if the system is to be a net producer of energy. In an ideal free economic system, the energy payback performance of a solar cell system or any other power plant would be adequately reflected in the cost of the system. In actuality, it is necessary for technologists and policy makers to consider the energy payback separately from the cost of energy produced whenever the government regulates or subsidizes some components of the energy industry.

When comparing different photovoltaic systems, one may consider their relative *tolerance to ambient conditions,* such as temperature, moisture in the air, and even the discoloring effect of sunlight itself on the panel covering, as these factors can shorten system life and increase the cost of the energy produced. Designs that are candidates for large-scale use must be *capable of being manufactured economically in large quantities.* The *availability of materials* used in such cells must also be examined, along with all the *environmental impacts* associated with manufacture, use, and eventual disposal of these cells.

In this chapter we shall consider many of these characteristics as we discuss two types of cells—single-crystal Si cells and thin-film CdS/Cu_2S cells, operating in AM1 sunlight. We will here also consider the monetary and energy costs associated with various processes, such as those for purifying silicon. In Chaps. 9 and 10 we return to these topics while discussing less conventional fabrication processes and materials that appear promising for making low-cost cells for widespread terrestrial deployment.

The properties of a semiconductor such as silicon depend upon the processes used in its manufacture. Of greatest importance is the degree of crystalline perfection exhibited by the final product. In this section we shall briefly discuss this characteristic, and then turn to the processing techniques commonly used to form semiconductors and make them into functioning solar cells.

Table 4.1 lists properties of single-crystal silicon and cadmium sulfide. Although the electrical properties of these semiconductors are of primary interest, the other properties may be important in cell design. One must know the index of refraction in order to design a proper antireflection coating. Mechanical properties determine the stresses the cells can withstand when they are encapsulated, shipped, and installed. Thermal properties are important for concentrator cells. (Properties of many other semiconductors that are candidates for advanced cell designs are given in Chap. 10.)

We referred in Chap. 1 to solids having different degrees of crystalline perfection, as illustrated in Fig. 4.2. In a perfect single crystal, the arrangement of atoms around any one lattice point is repeated exactly at all other lattice points. Real crystals may approach such perfection. In a polycrystalline solid, long-range order of this sort exists *within* individual grains having diameters ranging from a fraction of a micron to millimeters. The term *semicrystal* used in Fig. 1.2 refers to polycrystalline materials with grain sizes at the very high end of this range. The grains are more or less randomly oriented with respect to each other, and thus where they meet the neighboring atoms do not "fit" correctly, resulting in local strains and distortions at the grain boundaries. In amorphous solids there is at most only short-range order that extends over just a few atoms at each location.

Until very recently silicon solar cells have been made from single-crystal material, to exploit its excellent electrical properties. The carrier mobilities

Table 4.1 Some Properties of Silicon and Crystalline Cadmium Sulfide

Property	Semiconductor	
	Si	CdS
Electrical		
Energy gap:		
E_g (eV)	1.12	2.6
Type	indirect	direct
Drift mobility:		
μ_n (cm^2/V·s)	1350	340
μ_p (cm^2/V·s)	480	50
Diffusion constant		
D_n (cm^2/s)	38	9
D_p (cm^2/s)	12	1.4
Intrinsic carrier concentration: n_i (cm^{-3})	1.6×10^{10}	
Relative dielectric permittivity (static) ϵ_r	11.7	8.9
Mechanical		
Crystal structure	cubic diamond	hexagonal
Density: ρ(kg/m^3)	2.238×10^3	4.82×10^3
Fracture strength (MN/m^2)	≈ 220	
(lb/in^2)	$\approx 31{,}000$	
Optical		
Index of refraction:	3.44	2.5
Thermal		
Thermal conductivity (W/cm·°C)	1.45	
Specific heat per unit volume (J/kg·°C)	7×10^{-4}	
Coefficient of linear thermal expansion (1/°C)	2.6×10^{-6}	
Melting point (°C)	1420	

are high, and there are no grain boundaries and few defects to act as sites for enhanced recombination of photo-generated electrons and holes. Recombination at defects reduces the minority-carrier lifetime and hence lowers the cell efficiency.

The minority-carrier lifetime is one of the most important electrical characteristics of the semiconductor in a solar cell, but it is not an intrinsic property of the semiconductor used. Rather, it can vary from one sample to another of a given semiconductor, and it can also change during the pro-

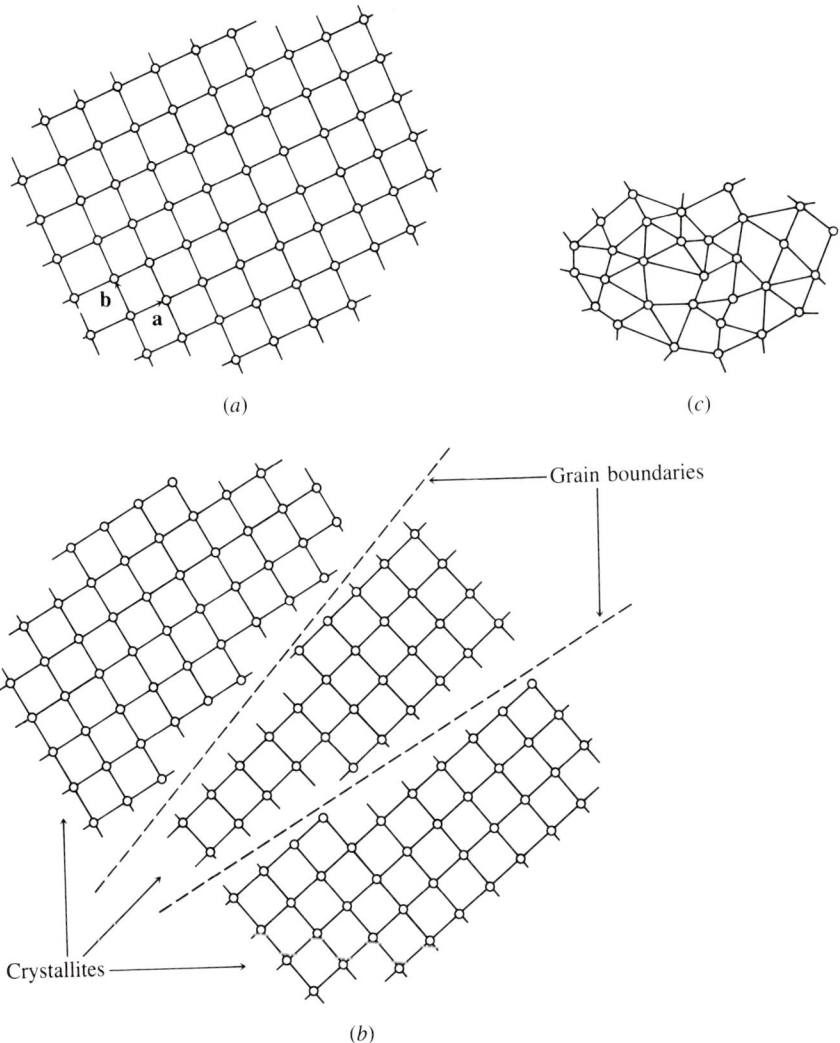

Figure 4.2 Portions of single-crystal, polycrystalline, and amorphous solids. (*a*) In a single crystal, atoms are regularly spaced throughout, and each atom can be reached from any other like atom by a translation that is the sum of integral multiples of primitive translation vectors such as **a** and **b**. (*b*) There is local order and translational symmetry in each crystallite of a polycrystalline solid, but the small crystallites may be differently oriented with respect to one another, meeting at their common borders to form grain boundaries. (*c*) There is no discernible local or long-range order in the amorphous solid, of which ordinary glass is the most familiar example.

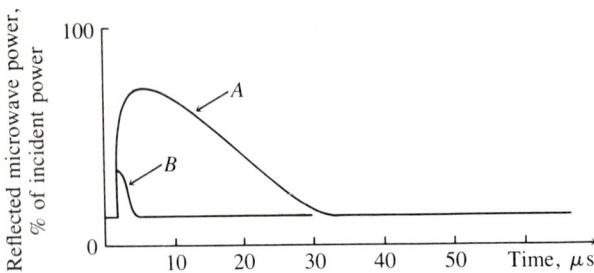

Figure 4.3 Oscilloscope displays of microwave power reflected by a silicon wafer when an intense flash of light creates electrons and holes that later recombine. Trace A is for an oxidized wafer, where recombination lifetime, indicated by the rate of decay of the pulse of reflected microwave power, is about 15–20 µs. Trace B, obtained with a bare, heavily ion-implanted silicon wafer, shows a very short lifetime.

cessing of a cell. This can be illustrated strikingly with the use of a microwave method to determine the lifetime.

In the microwave method, an oscilloscope displays the microwave energy reflected from a semiconductor wafer as a brief flash of light creates electron–hole pairs in the wafer. More microwave energy is reflected when wafer conductivity is high because of the light-produced carriers. As the carriers recombine, the conductivity and the microwave reflectivity fall. Figure 4.3 shows such a display of reflected microwave power from a carefully oxidized silicon wafer having a carrier lifetime around 15 or 20 µs (trace A). Trace B in Fig. 4.3 was made after ion implantation of the wafer—bombarding it with fast-moving ions, as described in the box in this chapter. The lifetime has decreased drastically because of the heavy damage caused by the impact of the high-energy ions near the surface of the wafer. This damage can be healed and the lifetime returned to nearly its initial value by heating the wafer in an inert atmosphere so that interstitial atoms can move into regular lattice sites. Similar reductions of lifetime occur when one etches away an oxide coating on a silicon wafer, or sandblasts the surface of a wafer, creating in both cases recombination centers at the surface at which electrons and holes can recombine.

This example shows how processing steps can affect material properties and hence cell performance. We now will consider the processes used to manufacture conventional silicon cells for terrestrial use.

4.2 CONVENTIONAL SILICON CELL PROCESSING

The four major steps in making silicon cells by presently conventional means are *forming crystals, doping, providing electrodes and an antireflection coat-*

ing, and *making a module* of interconnected cells, protected from chemical and physical attack from the atmosphere in which it is to operate.

Forming Single-Crystal Wafers

The objective is to produce large single-crystal wafers of Si three or more inches across, which have the following characteristics:

- Small number of defects per unit area.
- Controlled low concentrations of impurities (typically at some stage in the processing there will be only a few impurity atoms per billion host atoms of Si).
- Final thickness from 50 to 300 μm.
- Usually, one surface smooth enough for photolithography.

The conventional approach has been to purify the raw materials, grow single crystals by processes involving melting and slow solidification in carefully controlled temperature gradients, and then saw the resulting single-crystal ingot into slices that are polished as a final step. The chief contributors to cost and energy use in manufacture have been the purification and crystal-growth processes. Another costly feature of conventional processing is kerf loss—loss of up to a third of the single-crystal material in the form of "sawdust" produced when the ingots are cut into wafers. By way of a preview, note that processing improvements aimed at lowering costs (described in Chap. 9) are connected with improved methods of purification and ways of forming reasonably perfect crystals having nearly the final thickness and so not requiring wafering.

A typical conventional purification and crystal-growth process for making semiconductor grade silicon is summarized in Fig. 4.4. Quartzite pebbles—sand—are mixed with a source of carbon and reduced in a high-temperature electric-arc furnace to produce a relatively impure molten silicon, which is drained off and cooled rapidly in a boat, forming metallurgical-grade polycrystalline silicon (MG–Si). Reaction with HCl produces trichlorosilane gas, which is condensed and fractionally distilled to yield a much purer product indicated in Fig. 4.4 as semiconductor-grade (SeG) trichlorosilane. Reaction with hydrogen in a chamber containing a silicon substrate heated electrically to 1000 to 1200°C results in deposition of SeG–Si on the substrate.

To form single-crystal and still purer silicon, either the Czochralski (CZ) or the float-zone (FZ) solidification process is employed. In the CZ process, a small oriented piece of single-crystal silicon used as a seed is put in contact with the surface of the molten silicon in a crucible which is slowly rotated. Silicon from the melt freezes on the seed, which is slowly raised, resulting in the formation of a cylindrical single-crystal ingot from 3 to 5 inches in

78 MATERIALS AND PROCESSING

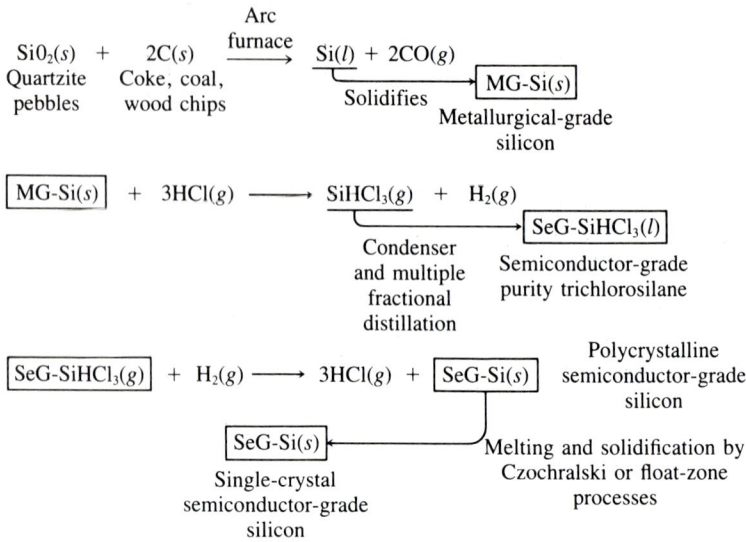

Figure 4.4 Conventional processes for making extremely pure, semiconductor-grade single-crystal silicon from sand. The symbols (g), (l), and (s) denote gaseous, liquid, and solid states, respectively.

diameter and up to 1 meter long. Some purification also occurs during CZ growth. In the float-zone method, a molten zone is passed through a relatively pure silicon ingot, causing a redistribution of impurities much as in the case of CZ growth. Because impurity concentrations in the liquid are generally higher than in the solid, passage of thin molten zones through an ingot having a 0.01 percent impurity content causes impurities to be swept to one end of the ingot, leaving in the middle a region having an impurity content as small as one part in 10^{10}. Finally, the cylindrical ingot is cut with saws into round wafers about 250 μm thick. Growth rates of only a few centimeters per hour are typical for CZ and FZ processes, and one goal of solar cell research and development is to find ways to make rapidly grown thin silicon sheets that do not require sawing.

Before following the wafers through the later steps of cell and array fabrication, let us consider the cost of this processing in terms of energy, money, and material. Many of the processing steps require volumes of material to be held at high temperature for many hours. This, together with the arc furnace reduction stage, requires a large energy input. At each stage of the processing, raw materials are lost. Of the silicon that is in the sand that enters the process, only about 20 percent emerges as SeG-Si, and about one-third of that is lost in the form of sawdust in the wafering step. Other materials are lost during the processing as well. The starting materials cost only about $0.21/kg, while the MG-Si is worth about $84/kg (1980 dollars) because of the labor, energy, and supplies consumed and the equipment used in its production. The 1986 DOE goal of $0.70/$W_{pk}$ cells can be reached with Si only if the cost of MG-Si can be lowered to about $14/kg.

The dilemma of conventional processing for Si cells can also be expressed in terms of energy payback time. At each stage of cell manufacture, energy is expended in three forms:

1. Direct energy, such as the heat needed to liquify polycrystalline MG–Si for single-crystal growth.
2. Indirect energy used in the mining, transport, and manufacture of materials utilized, such as the HCl for the refinement stage.
3. Equipment and overhead energy embodied in the equipment used in the refining plant and energy expended in lighting, heating, and cooling the factory.

The individual and cumulative energy payback times for 12.5 percent efficient Si cells (Lindmayer et al., 1977) are shown in Fig. 4.5. The total payback time is 6.4 years, meaning that the cells would have to operate that long to produce energy equal to that used in their manufacture. The largest single components in this requirement are the direct energy used in Si refinement and the indirect energy used in panel building, accounting respectively for 2.6 and 1.0 years of the total. In Chap. 9 we shall describe an improved refinement process for which the energy payback time for reduction of sand to silicon and refinement together is reduced to only 0.85 years.

Doping

The objective is to create planar regions having different concentrations of impurities so as to form a *pn* junction. Thus in *n* on-*p* Si cells, one may start with a *p*-type wafer containing boron as its chief impurity, and produce a very thin *n*-type layer near its surface by introducing phosphorus impurities into the wafer.

Impurities may be added intentionally to a molten semiconductor before solidification. The traditional method of forming a layer of different conductivity on the surface of such a wafer has been thermal diffusion of dopant atoms present in a gaseous molecule in a high temperature furnace in which the carefully cleaned wafers are placed as described in the box, Making an Integrated Circuit, earlier in this chapter.

Usually a two-step diffusion process is employed. In the *predeposition step* the desired impurity, transported by a carrier gas to the hot semiconductor surface, is diffused to a depth of a few tenths of a micron. For example,

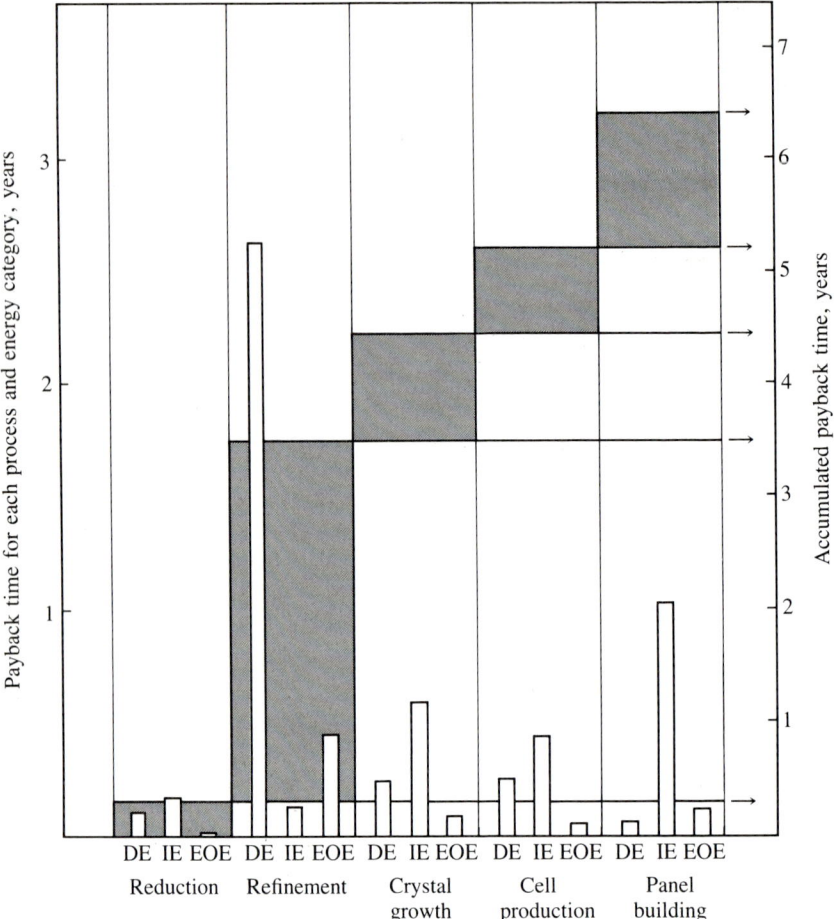

Figure 4.5 Energy payback times for fabricating 12.5-percent efficient AM1 silicon solar panels by conventional processes. Individual payback times for each stage, at left, and accumulated times, at right, are shown. Symbols: DE = direct energy; IE = indirect energy used in making materials expended in each stage; EOE = equipment and overhead energy. *(After Lindmayer et al., 1977).*

to dope Si with P, one may bubble nitrogen gas through liquid $POCl_3$ into a furnace held at from 800°C to 1100°C for anywhere from minutes to an hour or so. Alternatively, a solid oxide such as P_2O_5 might be heated for producing gaseous P_2O_5 which will react at the hot Si surface to form P and solid SiO_2. In the *drive-in step*, the semiconductor is simply heated and the predeposited dopant atoms diffuse into the semiconductor to a greater depth, up to about 1 μm.

Electrodes and Antireflection Coatings

In *pn*-junction solar cells, ohmic contacts are typically made with a uniform metallic electrode over the back surface and a fingerlike front electrode that permits light to pass between the opaque fingers. A transparent antireflection coating promotes transmission of light into the cell by reducing reflection, as discussed in Chap. 3. Some cells employ texturing of the semiconductor to reduce reflection.

Electrodes for conventional *pn*-junction cells must make ohmic contact to the underlying semiconductor; they should be highly conducting and should bond well by soldering to contact wires or conducting supports. Aluminum, gold, silver, and copper are conductive and solder well, but for ohmic contacting a heat treatment and additional metal layers between the metal and the semiconductor are often necessary. Furthermore, a metallic barrier layer may be needed to prevent diffusion of the electrode metal into the semiconductor where it might short-circuit the *pn* junction. As an example, an aluminum electrode deposited by evaporation onto the bare surface of a shallow-junction silicon *pn* diode requires a 5-minute heat treatment, typically at around 450°C, in forming gas (90 percent nitrogen, 10 percent hydrogen) or argon to make ohmic contact. During this treatment, aluminum might diffuse far enough in some regions to form conductive spikes that penetrate the junction, rendering the diode useless. To prevent this, one may deposit an Al–Si mixture with 5 percent Si instead, or evaporate a flash—a thin layer 100 to 200 Å thick—of a metal such as titanium or chromium which functions as a barrier to the aluminum that is evaporated over it.

For large-area flat-plate cells, electrodes may be deposited by any of several processes: evaporation of metal from an electrically heated filament or boat in a vacuum chamber, through an apertured mask; silk-screening onto the cell of a conductive paste such as an epoxy filled with metal powder with subsequent firing; and evaporation through a patterned layer made photolithographically in photoresist. This last process is used chiefly when the electrode pattern involves widths less than 100 μm, as may occur in concentrator cells. In such cells the currents are higher than in flat-plate array cells and electrode lines must be placed closer together on the cell to avoid large resistive losses. Once an electrode has been formed, it can be thickened by electroplating to reduce its series resistance.

As with veins in the leaf of a growing plant, which carry the products of photosynthesis away from individual cells (Fig. 4.6), the dimensions of the fingerlike front electrode on a solar cell are decided by an optimization process involving a compromise between reducing power losses due to series resistance and lowering the efficiency by making an overly dense, opaque electrode. Transparent conductive coatings may be used, as described further in Chap. 10. Cost and encapsulation considerations affect the thickness of the

Figure 4.6 The pattern of bright fingerlike electrodes on a silicon solar cell is optimized to collect current efficiently without preventing much sunlight from reaching the underlying semiconductor. Note similarity with pattern of veins in leaf next to it.

electrode used. Simple calculation shows that a strip of aluminum 50 μm wide and 15,000 Å thick (bulk resistivity of Al is 2.7×10^{-6} Ω·cm) has a resistance of 3.6 Ω for each centimeter of length. Also important is the spacing between metal electrode strips. For 0.5-cm spacing, the diffused layer over the same 1-cm length of the cell would have a series resistance of 5 Ω. Reduction of the resistivity of the doped semiconductor layer between electrodes is only effective to a limited extent because doping that region too heavily has been found to degrade cell efficiencies (Sec. 3.7). Most commercial Si cells have electrodes roughly 50 μm thick, around 100 μm wide, and spaced by about 0.5 cm over the cell. Much finer, closely spaced electrodes are used on concentrator cells, illustrated in Chap. 7.

Just as with a coated camera lens, reflections are reduced by putting on antireflection (AR) coatings consisting of layers having indices of refraction between that of the semiconductor and the air. The simplest AR coating is a single layer whose thickness is one-quarter wavelength of light in the coating and whose index of refraction n_c is the geometrical mean of the indices of the two media contacting it, $n_c = (n_s)^{1/2}$, or 1.84 for Si. Though no material having exactly this index is known, some oxides and fluorides approximating it do exist (see Table 4.2). For example, over the whole usable solar spectrum, 90 percent of the light is transmitted through a 600-Å coating of Ta_2O_5 on Si.

Antireflection coatings are applied primarily by evaporation; the process can be continuous if the vacuum system has suitable pumped entrance and exit

Table 4.2 Indices of Refraction of Materials Usable as Anti-Reflection Coatings in Solar Cells
Indices are typical values for wavelengths near the middle of the solar spectrum

Material	Index of refraction
Al_2O_3	1.77
Glasses	1.5–1.7
MgO	1.74
SiO	1.5–1.6
SiO_2	1.46
Ta_2O_5	2.2
TiO_2	2.5–2.6

ports. Other means of coating include spraying, dipping, and spinning-on liquids that dry and are then heated to densify them.

The simple quarter-wavelength coating can be improved upon in several ways. With multiple layers, having the lowest index layer toward the sun, the reflection can be kept low over more of the solar spectrum than with the single layer. The presence of so-called native oxides (oxides that simply form in air at room temperature) on the semiconductor material of the cell itself may need to be taken into consideration in the coating design. Chemical means exist for texturing the semiconductor surface and even for producing a gradually varying index in a cover glass put over the cell to greatly improve transmission into the cell. The textured surface results when a single-crystal semiconductor is treated with an orientation-dependent etch that attacks some crystal planes faster than it does others. For example, etching a Si wafer whose normal orientation is in the (100) or equivalent direction in heated (90 to 95°C) hydrazine (36 percent N_2H_4) produces small pyramids whose faces are (111) planes (Fig. 4.7). Light incident on these pyramids is partially transmitted and partially reflected at each contact with their surfaces as the light bounces its way down toward the bottom of the pyramids. As with a properly designed AR coating, the surface of a textured cell appears dark when viewed in white light.

Fabricating Arrays

Objectives of the final steps are to connect individual cells electrically and mechanically, and to protect them from degradation due to thermal, chemical, and physical causes during use. Since about one-third of the final cost of a commercial cell is incurred in this stage, it is not surprising that automated array assembly is envisioned for low-cost cell fabrication.

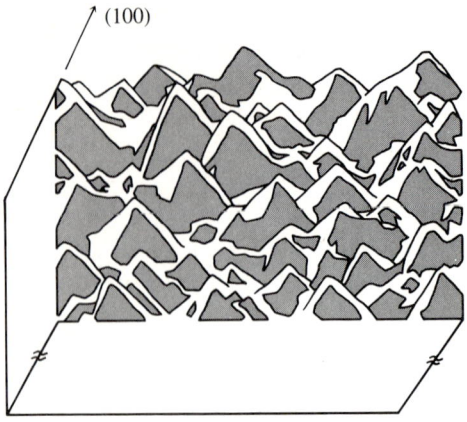

Figure 4.7 Textured surface of silicon that underwent orientation-dependent etching in hydrazine.

Conventionally, arrays have been made by electrically interconnecting individual cells, fastened to a plastic base, by wires or metal foil strips in a series-parallel arrangement. Finally, a glass or plastic sheet is fastened on the front in optical contact with the cells. Silicone-based potting compounds have been used to seal the cells for protection against degradation, particularly of their metallic electrodes, due to contact with moisture, salt, or reactive gases in the atmosphere.

Study of solar cell failures has shown that the encapsulant is responsible for many of the problems experienced. Failures have resulted from the separation of front covers from cells resulting in lowered cell efficiency because of the worsened optical contact between cover and cells, and from cracking of the encapsulant and subsequent penetration of moisture, and by discoloration of the covers. Best results usually are obtained with somewhat flexible panel assemblies which can adjust to strains produced during transport and installation of the array, and by temperature changes during actual use. Thus, potting compounds having a rubbery consistency tend to be favored, an example being silicone-based room-temperature-vulcanization (RTV) compounds, and resins that polymerize yet remain somewhat pliable. For protection against scratching due to abrasion in windstorms and sandstorms, glass has proven best. To avoid the discoloration noticed by desert travelers who find old bottles that have turned somewhat purple, glass with a low iron content is recommended. Among the plastics, polymethylmethacrylate has been found to retain its mechanical and optical properties well in 20-year exposure to actual AM1 sunlight in tests in the Arizona desert, whereas many other plastics weaken mechanically and discolor in just a few years in such light. The problem of concentrator cells is more difficult, because of the very high light intensity and the need for removing excess heat from the cells (see Chap. 5). Solar simulators exist for carrying out accelerated life tests on cells and complete arrays, but more must be learned in this area since the cells themselves should last many decades *provided* they are protected from attack by the ambient atmosphere.

4.3 PROCESSING CADMIUM SULFIDE CELLS

Totally different processes are used for making polycrystalline thin-film CdS/Cu_2S cells. Forming single crystals is not required, and the emphasis is on inexpensive continuous methods for making large-area modules that, although less efficient than the single-crystal Si cells, are nevertheless economical.

The conventional CdS cell (Fig. 4.8) is made by evaporating a 25-μm-thick polycrystalline CdS layer onto a 25-μm-thick copper or molybdenum foil which serves as the back contact. The foil is then dipped briefly in CuCl solution at slightly below 100°C and Cu atoms replace Cd atoms at the surface to form an approximately 0.2-μm-thick Cu_2S layer. This layer accounts for the cell's response to photons from its bandgap, 1.1 eV and up. A contact grid of Cu or Au is evaporated or pressed on the top, and the cell is encapsulated with a plastic, such as a mylar sheet held in place with epoxy. Heat treatments may be used to improve cell photoresponse.

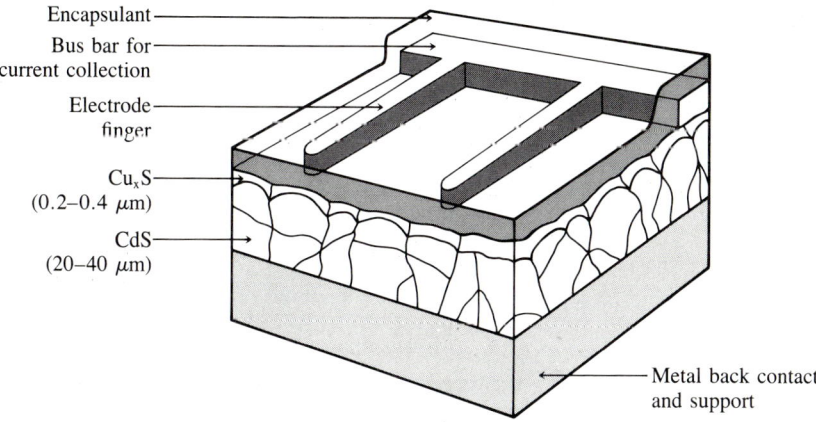

Figure 4.8 Cross-sectional view of polycrystalline cadmium sulfide (CdS/Cu_2S) solar cell.

A variety of other fabrication methods has been investigated, including use of a glass substrate coated with an optically transmitting and electrically conducting tin or indium oxide layer, and the use of spraying rather than evaporation to form the CdS layer (see Secs. 9.5 and 10.4).

Recombination at grain boundaries, leading to a lowering of the collection efficiency, and migration of Cu along grain boundaries into the cell, resulting in partial short-circuiting of the junction, are problems resulting directly from the polycrystalline structure of this cell. Additionally, the cells may suffer from oxidation of the Cu_2S to CuO and Cu_2O at temperatures above 60°C if air penetrates the encapsulation, and a light-activated change of Cu_2S to lower forms of Cu_xS has been identified as a problem in some of these cells.

Concerted work on the materials and the processing problems associated with these cells has resulted in steady increases in reported efficiencies and led to some optimism that such thin-film cells, having adequate operating lifetimes and possibly involving CdS with materials other than Cu_2S, will meet the 1986 cost goals (see Sec. 10.4).

4.4 ENVIRONMENTAL AND OTHER CONSIDERATIONS

So long as their costs are high, solar cells and photovoltaic (PV) systems will be used in very small quantities, and decisions about their design can be based strictly on technological grounds. As their costs fall and large numbers of PV systems are deployed, however, materials and processing techniques must be chosen only after careful consideration of additional factors—environmental impact, energy content of materials used in the systems, world supplies of the materials (discussed in Chap. 10), and the possible production rates that can be achieved in high-volume manufacture. One might add to this list, for less-developed countries, the possibility of local manufacture of cells and systems.

Environmental Effects

During normal operation, PV systems pose far fewer adverse environmental problems than do fossil-fueled or nuclear power plants. PV arrays do increase the absorption of solar energy locally and so could cause local changes of temperature and airflow, but no serious effects have been identified when large PV arrays have been simulated by areas of increased absorptance in computer models of global atmospheric behavior.

Some damage to plants, burrow-dwelling animals, and the fragile desert surface could occur as PV arrays are installed in desert regions. But more worrisome are two direct effects: (1) air contamination and possible human health hazards due to emissions associated with the mining, transport, and processing of ores to form PV arrays, and (2) biological effects of ingesting or inhaling toxic substances released if cells are on a building that burns accidentally.

One way to express the probable impact of the former type of emissions is to compare those from a PV system with those from a conventional power plant. Studies have shown (OTA, 1978) that routine emissions of SO_2, NO_x, CO, hydrocarbons, particulates, and waste solids in the manufacture and use of concentrator PV systems having 20 years life would be produced in from only 3 months to about 2 years of (manufacture and) operation of a coal-burning power plant, with scrubbers, having the same rated power output. Corresponding figures for flat-plate PV systems range from $3\frac{1}{4}$ months to as much as 9 years, depending on the technology chosen. These figures indicate that PV systems in normal operation are indeed relatively benign.

Quite hazardous chemicals are used in the manufacture of semiconductor devices, examples being PH_3 and BCl_3 used as sources of impurity atoms of P and B in doping, and of HCl and carcinogenic organic chemicals in cleaning processes. Safe procedures for using such substances have been developed for the integrated circuits industry, but since the volumes used will be orders of magnitude larger in a fully developed solar cell industry, these emissions will have to be carefully controlled.

Concern for the toxicity of the semiconductors in PV systems themselves is based largely on accidental dispersal as a result of fire. There are no toxic effects directly associated with Si. Cadmium is a heavy metal posing serious health hazards for humans, yet anticorrosion coatings of cadmium have been used routinely for years. If GaAs cells were involved in a serious fire, toxic and carcinogenic arsenic compounds such as As_2O_3 might be released into the atmosphere, contaminating it and also perhaps nearby water supplies. That GaAs is expected to be used only in systems where the sunlight is concentrated by a lens or focusing mirror is a further worry, since the cells could vaporize if their cooling systems failed. The hazard of high temperatures in concentrator systems is offset, however, by the small quantity of semiconductor material present. One should also not ignore the hazards of concentrated sunlight to rooftops, if the tracking machinery should go awry, and to eyesight if even a small fraction of the concentrated beam is reflected into human eyes.

Energy Content of Materials Used

The concept of energy payback time, already mentioned in connection with the semiconductors in the cells, also applies to the choice of materials for PV system supports, tracking structures if any, antireflection coatings and encapsulants. As Table 4.3 shows, the equivalent thermal energy required for a unit

Table 4.3 Energy Consumption in Materials Processing

Ton used here, 2000 lb. Estimated uncertainty of entries, ±15% except order-of-magnitude entry for plastics is same as average for inorganic chemicals [*Makhijani and Lichtenberg (1971)*]

Material	Energy per unit of production, kWh_T/ton
Titanium (rolled)	140,000
Aluminum (rolled)	66,000
Copper (rolled or hard drawn)	20,000
Steel (rolled)	11,700
Zinc	13,800
Lead	12,000
Glass (finished plate)	6,700
Paper (finished, average)	5,900
Inorganic chemicals (averaged)	2,400
Plastics (order-of-magnitude, assumed same as inorganic chemicals)	2,400
Cement	2,200
Coal	40
Sand and gravel	18

of production varies greatly from material to material. To these figures must be added energy required for transportation, which is much smaller for locally produced materials such as sand, gravel, and cement than for materials such as steel and aluminum. When one includes energy for transportation over typical distances, along with energy lost in the depreciation of production machinery, the total energy expenditure per ton of cement is only 2300 kWh_T/ton, whereas that for aluminum reaches about 67,000 kWh_T/ton. (The subscript T denotes thermal energy, where 1 $kWh_T = \frac{1}{3}$ kWh of electrical energy.) The low energy required for production and use of cement, glass, and plastics suggests that they be used wherever possible in PV systems.

Production Rates

Once the design for a cost-effective, energy-efficient, and environmentally benign solar cell is found, the challenge is to manufacture it in sufficient quantities and fast enough so it can meet increasing demands for electricity. The extremely high rates of production of cells and optical coatings and encapsulants required can be appreciated from the following example.

Example: Suppose that the consumption of electrical energy in the U.S. during the year 1980 was 2.2×10^{12} kWh, that the rate of consumption increased by 3 percent annually through the year 2000, and that we wish to meet 10 percent of the electricity demand in the year 2000 with 11 percent efficient solar cell systems. Assuming that the insolation is equal to that in El Paso, Texas (2000 kWh/m² per year), what area of cells would be required by the year 2000, and what would be the annual production rate in square meters if 5 percent of the total were to be manufactured each year?

$$\text{Area required} = \frac{(\text{total electrical energy in year 2000})(0.10)}{(\text{annual insolation}/m^2)(0.11)}$$

$$= \frac{(2.2 \times 10^{12})(1 + 0.03)^{20} (0.10)}{(2 \times 10^3)(0.11)}$$

$$= 1.8 \times 10^9 \text{ m}^2 \text{ (or about 700 mi}^2\text{, an area about 26 miles on a side)}$$

$$\text{Area/year} = 1.8 \times 10^9/20 = 9 \times 10^7 \text{ m}^2/\text{yr (about 35 mi}^2/\text{yr)}$$

Thus it is necessary to produce about 35 mi² of arrays each year—an area that is 5.9 mi on a side. This production rate is also impressive when put in terms of the tons per year of materials required. A 2-μm-thick CdS film having a density of 4.8 g/cm³ weighs about 20 tons per square mile, and a silicon wafer of standard thickness (0.010 in, or 250 μm) at 2.33 g/cm³ weighs about 1170 tons per square mile. Thus a production of 35 mi² per year would amount to about 40,000 tons of silicon per year if wafers of standard thickness were used, an amount that is about 30 times the present annual U.S. production of semiconductor grade silicon. Clearly it is advantageous to use thin cells, and we must employ production processes capable of high rates if photovoltaic systems are to contribute significant amounts of energy.

Fortunately, some industrial plants are capable of the required high rates of production (Mattox, 1975). For example, an air-to-air strip coater exists that can coat a 150-cm-wide plastic sheet with 500 Å of aluminum at a speed of 60 m/min, resulting in a throughput of 5×10^7 m²/year. Float-glass production in a single plant at a rate of 3×10^7 m² (10 mi²) per year has been achieved. Both electron-beam evaporation and sputter-deposition coating of large glass panels are routine, high-rate industrial processes. Electrodeposition is a low-cost process that is also a candidate for high-volume production. For example, plants exist that electrodeposit 1-μm-thick tin layers on strip steel at rates of 500 m/min at a cost of a few cents per square foot, for an annual production of about 1.9×10^8 m². The rates quoted are applicable in principle to production of PV systems because these processes can be used to make antireflection coatings, contacts and, in the case of thin-film PV cells, the semiconductor layers themselves.

4.5 SUMMARY

Terrestrial solar cells have been made commercially of single-crystal silicon—because of its excellent electrical properties and well-developed technology—and of thin-film CdS/Cu$_2$S—because of its low cost per unit area. To reduce the cost per watt of output, higher efficiency cell designs, and less energy-intensive and more automated processing is required. Adverse environmental impacts from silicon solar cell use are much less than for well-scrubbed coal-fired power plants. A number of industrial processes now used for other purposes exhibit production rates of the magnitudes required for significant solar cell and concentrator manufacture.

REFERENCES

Lindmayer, J., Wihl, M., and Scheinine, A. (1977), Energy Requirement for the Production of Silicon Solar Arrays, Report SX/111/3, Solarex Corp., Rockville, Md., October.

Makhijani, A. B., and Lichtenberg, A. J. (1971), An Assessment of Energy and Materials Utilization in the U.S.A., ERL-M310 (revised), Electronics Research Laboratory, University of California, Berkeley, September.

Mattox, D. M. (1975), Solar-Energy Materials Preparation Techniques, *J. Vac. Sci. Tech.*, Vol. 12, No. 5, 1023–1031, September.

OTA (1978), Application of Solar Technology to Today's Energy Needs, 2 vols., Office of Technology Assessment, Congress of the U.S. (June, September).

PROBLEMS

4.1 *Production of dopant required* For the production rate of silicon cells worked out in the Example in this chapter, determine the corresponding rate of production of phosphorus required to dope the thin front layer on the cells. Assume the front layer is doped n-type to about 10^{19} phosphorus atoms per cubic centimeter for a thickness of 1 μm, and determine the weight of phosphorus required per year. (The dopant is diffused in from the front surface of the silicon and distributed in a gaussian fashion, with a total number of dopant atoms per unit of cell area roughly equal to the value you will obtain assuming uniform distribution.)

4.2 *Need for flexible encapsulant* In use, a solar cell might experience ambient temperatures from -20 to $+40°C$ or more. Find the change due to thermal expansion of the diameter of a 10-cm-diameter silicon solar cell and the 10-cm length of an aluminum mounting plate that might be used to support the cell (assume a thermal expansion coefficient of $24 \times 10^{-6}/°C$ for aluminum and $2.6 \times 10^{-6}/°C$ for silicon).

4.3 *Mineral supply for CdS cells* The weight ratio of Cd to S in CdS is 3.5:1, and the available supply of S is much larger than that of Cd. Using the same assumptions as in the Example in this chapter, find the peak watts that can be generated by CdS cells if all of the world production of Cd in one year, 2×10^4 metric tons, and if all of the identified Cd mineral resources in the world, 1×10^6 metric tons, are used to make CdS solar cells.

4.4 *Silicon material cost* Referring to Fig. 4.4, silicon costs about 10¢/kg in the form of quartzite ore, $1/kg as MG–Si, $80/kg as purified polycrystalline Si, $250/kg as single crystal Si, and $400/kg as wafers after 30 percent sawing loss. If the wafers are 0.025 cm thick and the specific weight of silicon is 2.33 g/cm^3, approximately how many peak watts can be generated by cells made from 1 kg of Si wafers? Use your intuition and arbitrary guesses to compare the potential savings in $/W$_{pk}$ as the results of a cheaper method of making MG–Si, (for example, one that reduces the cost of this step by half), of purifying Si or making solar cells from less pure Si, of growing crystal or cutting crystal into wafers, of making solar cells with polycrystalline Si, and of making 10-μm-thick (amorphous) Si solar cells. Try to be quantitative, though arbitrary. Many specific cost reduction methods will be discussed in Chaps. 9 and 10.

CHAPTER FIVE
CONCENTRATION OF SUNLIGHT

Technical and economic characteristics of optical concentrators and cells designed for use with them

CHAPTER SIX
POWER CONDITIONING AND ENERGY STORAGE

Electronic systems for optimizing power output and overall usefulness of solar cells

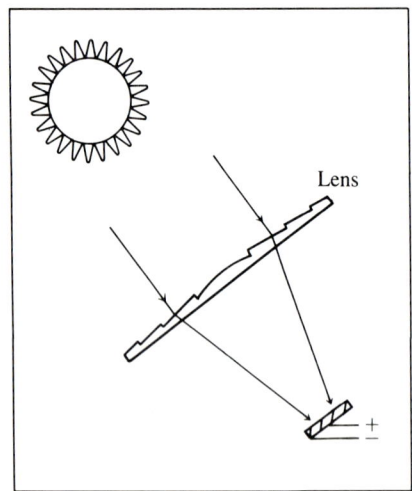

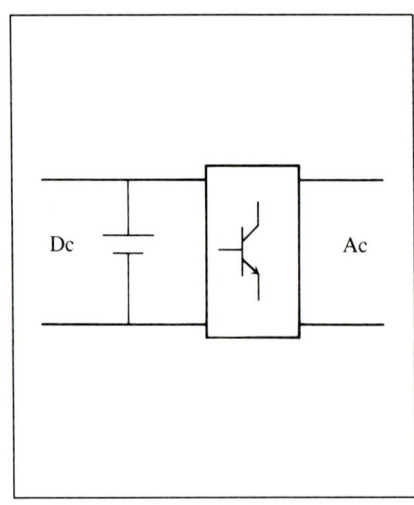

PART TWO

CONCENTRATORS AND COMPLETE PHOTOVOLTAIC SYSTEMS

CHAPTER SEVEN
CHARACTERISTICS OF OPERATING CELLS AND SYSTEMS

Applications and operating characteristics of actual photovoltaic systems

CHAPTER EIGHT
ECONOMICS OF SOLAR CELL SYSTEMS

Cost analyses and other factors affecting solar cell use in industrialized and developing countries

CHAPTER

FIVE

CONCENTRATION OF SUNLIGHT

CHAPTER OUTLINE

5.1 SOLAR CONCENTRATORS
BOX: *f*-NUMBER OF A CAMERA LENS
BOX: THE EQUATION OF TIME
5.2 ECONOMICS OF CONCENTRATOR PHOTOVOLTAIC SYSTEMS
5.5 SUMMARY
REFERENCES
PROBLEMS

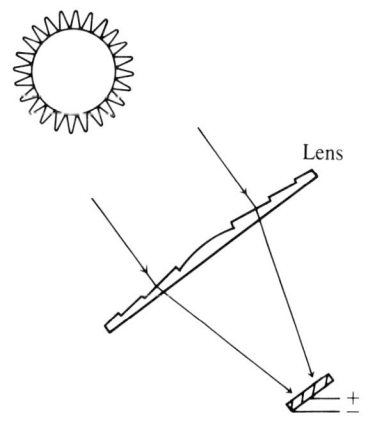

The intensity of solar radiation, about 1 kW/m² at noon on a clear summer day, the cost of each unit area of solar cells, and the cell efficiency together determine the minimum capital cost per watt of solar cell capacity. The cost of electricity generated from solar cells is high compared with that from conventional sources. One possible solution to the cost problem is to replace expensive solar cell area with low cost concentrator area and still derive about the same total power output. The economics of photovoltaic systems using sunlight concentration will be discussed in this chapter, with consideration of the relative cost advantage, the optical concentrators, and the need for tracking the sun.

A less obvious advantage of the concentrator photovoltaic system is that the conversion efficiencies of solar cells can be higher under concentrated sunlight. Furthermore, in a concentrator photovoltaic system, the cost of the solar cells is often only a small portion of the system cost, and relatively more extensive engineering efforts and costly cell manufacture technologies can be applied in order to bring about the high efficiencies. The reasons for the higher efficiencies and some other special considerations of concentrator solar cells will be presented.

5.1 SOLAR CONCENTRATORS

Sunlight concentration is not a new concept. In the history of solar energy applications (Meinel, 1977), particularly those that caught the public's attention, solar concentrators were used more often than not. Archimedes is said to have burned the invading Roman fleet of Marcellus in 212 B.C. by means of concentrated sunlight. Solar concentrators have been used to melt materials and to power heat engines. In the seventeenth century, lenses or burning glasses were the dominant form of concentrator. In the eighteenth and nineteenth centuries, the most popular concentrators were cone-shaped or parabolic reflectors. In the early twentieth century, many large solar energy projects used parabolic trough reflectors.

Today's concentrators for photovoltaic systems still have these basic forms. Bulk lenses, however, have been ruled out because of weight and cost, and are usually replaced by thin fresnel lenses to be described later.

Basic Concentrator Concepts

In evaluating a solar concentrator, the following are the key parameters to consider.

Concentration ratio If the collecting area of the lens or the reflector is 100 times the area of the solar cell, the concentration ratio of the system is said to be 100. The symbol "100×" is often used to denote this fact. The implicit assumption is that the cell would be illuminated by light 100 times more intense than normal sunlight. In fact, the light intensity at the cell surface is almost certainly quite nonuniform because of the aberrations of the optical system. Furthermore, the average light intensity at the cell surface is likely to be less than 100 times the normal sunlight intensity because of the reflection, transmission, and scattering losses incurred along the paths of sun rays in the concentrator.

On the other hand, when a cell is said to have been tested under 100× AM1 illumination, the supposition is that it has been tested under uniform illumination equivalent to 100 times the intensity of AM1 sunlight.

Imaging and nonimaging concentrators If the solar cell is placed in the focal plane of a lens (or a reflector), as shown in Fig. 5.1a, the concentrator is said to be an *imaging* concentrator, as the sun's image is formed on the cell surface. Figure 5.1b shows a concentrator that does not form an image of the sun. Such concentrators are called *nonimaging* (Welford, 1978). The distinctions between imaging and nonimaging concentrators are often vague because many "imaging" concentrators are of such poor optical quality that they don't form any clear images and the solar cell may be placed away from the "focal plane."

One-dimensional and two-dimensional concentration A parabolic dish, for example, provides two-dimensional (2-D) concentration; a parabolic trough (see Fig. 5.2) provides one-dimensional (1-D) concentration.

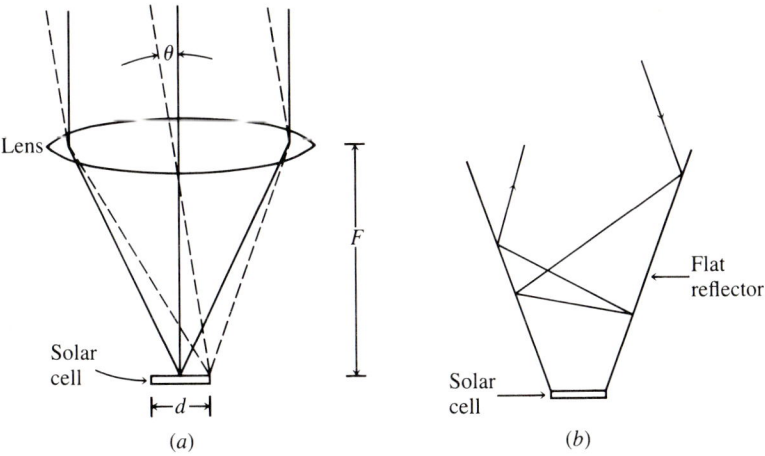

Figure 5.1 (a) Imaging concentrator. (b) Non-imaging concentrator. For either type, the acceptance angle decreases as the concentration ratio increases.

98 CONCENTRATION OF SUNLIGHT

Acceptance angle In an imaging concentrator, as Fig. 5.1a shows, only rays incident within θ from the axis of the system can be focused on the solar cell. θ is called the *acceptance angle*. Lens (or reflector) optics shows (Born and Wolf, 1970)

$$d \approx 2F \sin \theta \qquad (5.1.1)$$

The concentration ratio of the system is approximately D/d for 1-D concentration and $(D/d)^2$ for 2-D concentration. Therefore, the concentration ratio

$$C \approx \frac{1}{2f \sin \theta} \approx \frac{29}{f\theta} \qquad \text{1-D concentration} \qquad (5.1.2)$$

$$C \approx \left(\frac{29}{f\theta}\right)^2 \qquad \text{2-D concentration} \qquad (5.1.3)$$

where f, the f-number, is the ratio of the focal length F to the lens diameter D and $\sin \theta \approx \theta/58$ with θ in degrees. For a 2-D concentrator with $f = 1$ and $C = 100$, the acceptance angle is about 2.9°. Rays coming from outside this cone of acceptance will not be directed to the solar cell. It is difficult to manufacture concentrators with f much less than 1.

---

f-NUMBER OF A CAMERA LENS

Next to the lens of a camera there is an opaque screen, the aperture, with a round hole in the center. The diameter of the hole is adjustable and determines the effective diameter of the lens. For a given lens, the focal length F is fixed. Since the f-number is defined as

$$f\text{-number} = \frac{\text{focal length of lens}}{\text{effective diameter of lens}} = F/D$$

the f-number is inversely proportional to the diameter of the aperture. The amount of light entering the aperture is proportional to the area of the aperture, and thus to D^2 or $1/f^2$; therefore,

f-number	1.4	2	2.8	4	5.6	8	11
Relative amount of light entering aperture	1	1/2	1/4	1/8	1/16	1/32	1/64

This table explains the odd-looking f-numbers (e.g., 5.6) that one encounters and why photographers who want to expose film to the same amount of light energy, increase by a factor of two the length of time the shutter is open when they reduce the diameter of the aperture by one f-number (e.g., from f2.8 to f4).

For a nonimaging concentrator, Liouville's theorem (Born and Wolf, 1970) states that the 1-D concentration ratio is limited to

$$C \leq \frac{1}{\sin \theta} \approx \frac{58}{\theta} \qquad (5.1.4)$$

Thus, there is an upper limit to concentration ratio for a given acceptance angle, and vice versa. Figure 5.1*b* illustrates the fact that rays coming outside the acceptance angle of a nonimaging concentrator may not reach the solar cell.

Solar tracking Earlier we saw the example of a 100× concentrator that has an acceptance angle of 2.9°. At the rate of motion of about 15° per hour (Chap. 2), the sun would sweep across the field of view of this concentrator (5.8°) in about 23 min. Clearly, a tracking mechanism will be needed to direct this concentrator toward the sun throughout the day. The tracking mechanism cannot be controlled precisely by a clock unless correction for the "Equation of Time" is provided (see Box). More interesting is the concept of automatic

---

THE EQUATION OF TIME

Sundials often include on them corrections (see photograph) for the small variations that occur through the year in the speed of the sun's apparent motion through the sky.

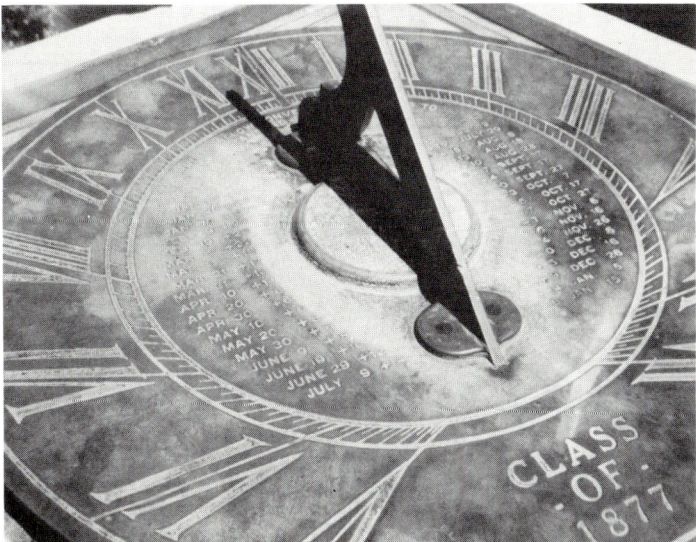

This sundial has stamped on it the number of minutes to be added to or subtracted from its indicated time on specific dates to correct for the varying speed of the sun's motion through the sky.

The average angular speed of the sun is 15° per hour. But if one observes the actual motion, one finds that the time of day at which the sun is at its highest point locally varies by as much as 25 min through the year. The so-called equation of time is defined as the time the mean sun, assumed moving at 15° per hour, is at the zenith minus the corresponding time for the actual sun.

The equation of time, plotted below as curve *C*, is composed of two main parts. One results from the variation of the earth's orbital speed owing to the change in earth-sun distance during the year. The earth's orbital speed is largest in January and smallest in June, causing the actual sun to arrive at its zenith earlier than expected during the first half of the year, and later in the second half (curve *A*). The second major effect arises from the tilt of the earth's axis. This correction is zero at the vernal and autumnal equinoxes, and at the summer and winter solstices. At other times of year, the tilt which causes the declination to change greatly also causes a small change in the time the sun is at the zenith (curve *B*).

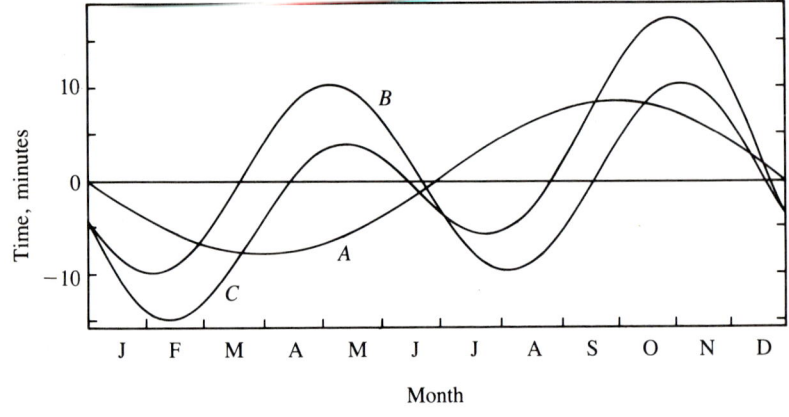

The correction given by the equation of time must be taken into account if one attempts to drive a tracking collector for solar cells by a clockwork mechanism.

solar tracking with the help of simple split-field photodiodes and electronics. To direct the concentrator during cloudy periods and to reset at the end of the day would probably require some artificial intelligence such as can be provided by a microprocessor. A few tracking mechanisms are commercially available (see Appendix 8). For some lower-concentration-ratio, one-dimensional concentrators, manual daily or even seasonal adjustments are satisfactory.

The highest possible concentration ratio in principle is set by the finite size of the solar disk, which requires an acceptance angle of 0.27° from the concentrator. From Eq. (5.1.3), assuming $f = 1$, a 2-D concentration ratio of 11,500 would be possible. In reality, the imperfections of the optical components and the inaccuracies of the tracking mechanism would probably set a lower limit of about 1° to the acceptance angle. This in turn sets the upper limit of concentration ratio at about 1000.

Types of Solar Concentrators

We shall now examine the many different types of solar concentrators that have been considered for solar cell systems. Intensive development efforts have improved the optical efficiencies of concentrators to the 80 percent range, verified the expected economic advantage of concentrator systems, and narrowed the choices of concentrators. Boes (1980) has reviewed the solar cell concentrator system development programs in the United States and pointed out that the parabolic trough and the fresnel lens concentrators are by far the most popular types of concentrators. These programs tend to emphasize modules of a few-hundred-watt size with concentration ratios between 25 and 100. The upper limit of concentration ratio seems to be determined by the ready availability of suitable solar cells.

Parabolic troughs Parabolic troughs (Fig. 5.2a) are one-dimensional concentrators with single-axis tracking. The length of the trough lies in the north–south direction, either on a horizontal plane, or tilted toward the south (for northern hemisphere locations, as mentioned in Chap. 2). The tilt angle, furthermore, can be fixed or adjusted daily or seasonally.

The reflective materials are usually silvered glass, polished aluminum, or aluminized flexible film. The latter two materials are relatively cheap, but the reflectivity over the spectrum of interest is only about 85 percent with a sharp dip at 0.8 μm wavelength. The reflectance of silvered glass can be as high as 90 to 95 percent. The necessary curved shape can either be achieved by thermal sagging or mechanical bending.

A specific example of parabolic trough concentrators is the design used in the 240 kW photovoltaic demonstration project at Mississippi County Community College. Developed by Solar Kinetics, the concentrator measures 7 ft × 20 ft (2.1 m × 6.1 m) with the long axis aligned north–south horizontal. It has a concentration ratio of 42 and uses both aluminized film and silvered glass in different modules.

Another example is a General Electric design to be used in the 100 kW photovoltaic/thermal system at Sea World, Orlando, Florida. It measures 7 ft × 30 ft (2.1 m × 9.1 m) and has a concentration ratio of 34. The reflector is made of aluminized plastic film. The trough is mounted on a turntable with the long axis tilted. Changing the tilt angle and rotation of the turntable provides the tracking.

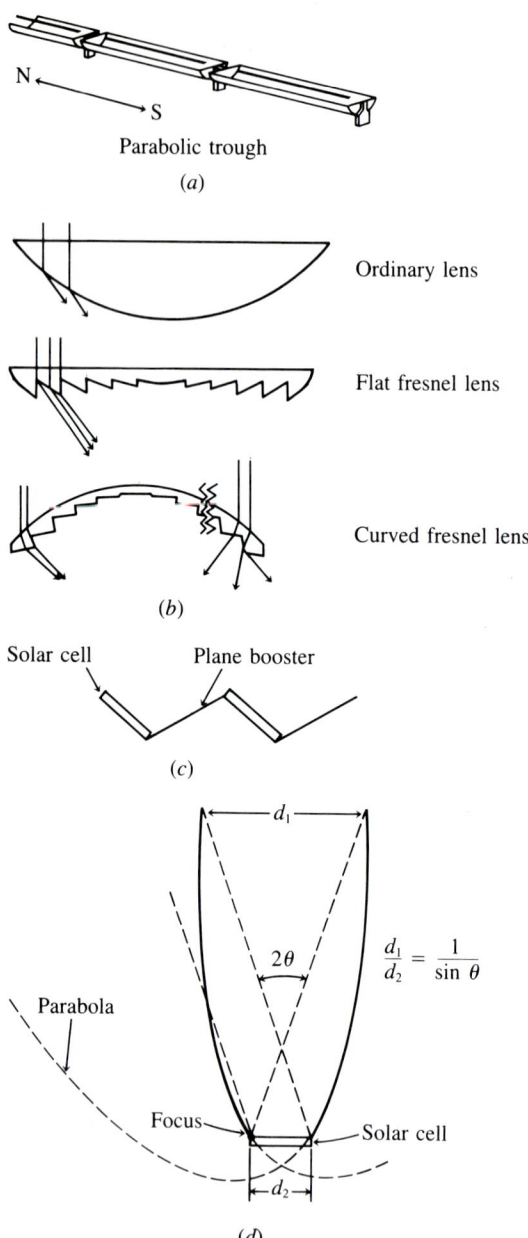

Figure 5.2 Types of concentrators: (*a*) parabolic trough, (*b*) fresnel lens, (*c*) plane booster, (*d*) compound parabolic concentrator, (*e*) parabolic dish, (*f*) central receiver concentrator, and (*g*) luminescent concentrator.

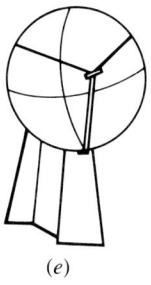
(e)

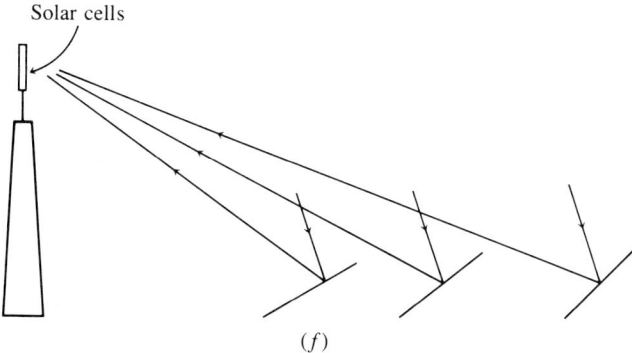
Solar cells

(f)

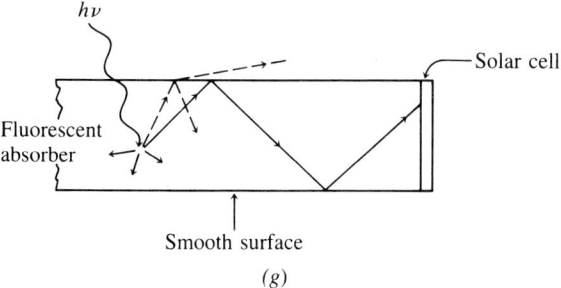
$h\nu$

Solar cell

Fluorescent absorber

Smooth surface

(g)

Figure 5.2 (*continued*)

Fresnel lens Figure 5.2b illustrates the derivation of the thin fresnel lens from a conventional lens. The sun rays are diffracted by a fresnel lens just as they would be by a conventional lens. The advantage of a fresnel lens is that it is much thinner and lighter than a conventional lens of the same diameter and f-number. Fresnel lenses are most often manufactured by injection molding of optical-quality plastic materials. It has been estimated that the sales price of cast acrylic fresnel lens (Swedlow, 1976) in 1980 dollars would be about \$60/m^2 at a production rate of 15,000 m^2 per year. At 100,000 m^2 per year, the price would be \$42/m^2 for 0.635-cm-thick lenses and \$33/m^2 for 0.254-cm-thick lenses. It should be mentioned that thin fresnel *mirrors* may be made similarly with an added step of aluminizing.

Complex ray-tracing programs have been used to determine the size and shape of the fresnel lens grooves in order to reduce reflection loss and aberration and to achieve uniform illumination at the cell. James and Williams (1978) have shown that a curved (spherical or cylindrical) fresnel lens can further reduce reflection loss as well as enhance the mechanical rigidity. The facets in a curved lens should ideally be slanted inward to avoid reflection loss, as shown on the right-hand end of the lens in Fig. 5.2b. This, however, presents difficulty for the removal from the mold after injection molding. The first example described below presents a clever solution to this problem.

E-Systems has developed a concentrator whose arched fresnel lens is formed by simply bending a molded flat fresnel lens. It measures 3 ft × 8 ft (0.9 m × 2.4 m). The long axis runs north–south and is tilted toward the south. The tilt angle is adjusted daily while the single-axis tracking about the long axis is continuous. This concentrator is said to be capable of concentration ratios over 1000. In the 27-kW experiment at the Dallas Airport, this concentrator will operate at 25× concentration.

Another example worth noting is the fresnel concentrator system used in the very large 350-kW Soleras village electrical power project installed in Riyadh under joint support by the United States and Saudi Arabian governments. Designed by Martin Marietta, Inc., the basic unit is a flat 1 ft × 4 ft (0.31 m × 1.2 m) molded lens that contains four 1 ft × 1 ft (0.31 m × 0.31 m) lenses. Sixty-four such units (see Fig. 7.5) are mounted on a pedestal equipped for 2-D tracking. The concentration ratio is about 40.

Flat-plate concentrator The simplest form of a flat-plate concentrator or booster is shown in Fig. 5.2c. The solar cell output can be enhanced by about 50 percent (Hill, 1977). This type of concentrator has been proposed for use with the space-satellite photovoltaic generation station (Sec. 12.4).

Compound parabolic concentrator Winston (1974) proposed an elegant realization of Liouville's limit [Eq. (5.1.4)]. This concentrator, also known as a *Winston collector,* not only provides a large acceptance angle but also allows all sun rays within the acceptance angle to reach the solar cell after at most one reflection. As shown in Fig. 5.2d, the 1-D concentrator is formed with two parabolic cusps that satisfy these two conditions: (1) the focus of one cusp must be placed on the other, and (2) a line connecting the focus of one parabola to the top of its cusp must be parallel to the axis of the other parabola. With these facts in mind it is easy to see that $d_2/d_1 = \sin\theta$

(Liouville's theorem) and that a sun ray within the acceptance angle θ reaches the solar cell after at most a single reflection.

Because of its rather large acceptance angle, this collector is used primarily as a nontracking concentrator with the length of the trough running east–west. At solstices, the sun stays within the acceptance angle of a 5× concentrator for up to 8.7 hours, and a 10× concentrator for 6.8 hours (Meinel, 1976).

The disadvantage of the compound parabolic concentrator is that it requires a large area of reflective surface in comparison with other concentrators. The upper portion of a full Winston collector, however, can be arbitrarily trimmed, resulting in a much shorter structure with only slightly reduced concentration ratio. The resultant structure is called a *truncated Winston collector*. A 2-D Winston collector can be generated by making a surface of revolution from the cross-section of the 1-D concentrator.

Parabolic dish As shown in Fig. 5.2*e*, parabolic dishes require two-axis tracking and provide very high concentration ratios.

Central receiver concentrator As shown in Fig. 5.2*f*, the total mirror-surface area in one module can collect from tens of kilowatts to tens of megawatts. This has lately been the preferred concentrator for solar-thermal-electricity generation but has never been seriously discussed for concentrator photovoltaic systems.

Luminescent concentrator As shown in Fig. 5.2*g*, this unusual concentrator is made of a liquid or solid plate containing fluorescent materials, such as an organic dye or Nd ions. Sunlight is absorbed by the fluorescent material, which re-emits light semi-isotropically. A good portion of the re-emitted light is trapped in the plate by total internal reflection until it reaches the edge of the plate where it is absorbed by the solar cell.

A severe limitation of this potentially inexpensive, nontracking concentrator is its low optical efficiency. At low concentrations of the fluorescent material, absorption of sunlight is far from complete, whereas high concentrations would cause a photon to go through multiple absorption/emission cycles before reaching the solar cell with less than unity quantum efficiency at each cycle. Many ingenious improvements have been suggested over the recent years (Rapp, 1978) but the optical efficiency is still only about 20 percent.

Holographic concentrator A hologram is basically a large diffraction grating that modifies the wavefronts of light waves in a predetermined and usually complex way. Its ability to form three-dimensional images is well known. A hologram can also perform simpler tasks such as diffracting or focusing a light beam. Since a hologram's operation is based on diffraction and is thus sensitive to the wavelength, a holographic concentrator cannot collect sunlight over the entire solar spectrum unless multiple holograms are used. The wavelength sensitivity, on the other hand, can be exploited for spectrum-splitting systems (see Chap. 12, Sec. 12.1). By using multiple holograms, the need for solar tracking can be eliminated (Ludman, 1982). Holographic concentrators are more expensive than fresnel lenses and their ability to stand long-term UV exposure is questionable.

5.2 ECONOMICS OF CONCENTRATOR PHOTOVOLTAIC SYSTEMS

As stated at the beginning of this chapter, the per-unit area cost of solar cells seems to be the largest and most variable cost item in a photovoltaic system. We shall now compare the maximum acceptable unit-area cell costs of a flat-plate and a concentrator system. The cost of electricity generated by solar cells can be estimated from the following equation:

$$\text{Cost/kWh} = \frac{\text{cost of plant} \cdot \text{amortization rate}}{\text{kWh produced each year}} + \text{operating cost} \quad (5.2.1)$$

For a relative comparison, we shall use the following numbers, which are believed to be reasonable.

Acceptable cost per kWh: $0.06
Cost of plant excluding cells
   Direct: $170 per peak kW for power conditioning, buildings, and battery storage for half day, and $30/m² for flat-plate system for land, support structure, etc., or $70/m² for concentrator system for land, concentrator, tracking structure, etc.
   Indirect: 40 percent of direct cost for architect and engineer fees, interest during construction, etc.
Amortization rate: 12 percent for interest and principal payments
Solar insolation: 4.4 kWh/day/m² (Chap. 2)
Operating cost: $0.005/kWh

Let C be the maximum acceptable cell cost ($/m²), X the concentration ratio, and η the cell conversion efficiency. Considering 1 m² of collection area, Eq. (5.3.1) becomes

$$0.06 = \frac{C + 30 + 170 \cdot \eta}{4.4 \cdot 365 \cdot \eta} \cdot 1.4 \cdot 0.12 + 0.005 \quad (5.2.2)$$

for a flat-plate system, or

$$0.06 = \frac{C/X + 70 + 170 \cdot \eta}{4.4 \cdot 365 \cdot \eta} \cdot 1.4 \cdot 0.12 + 0.005 \quad (5.2.3)$$

for a concentrator system. The maximum acceptable costs per m² of cells are listed in Table 5.1. The table leads to several interesting conclusions:

1. Even if cells cost nothing, they must have some minimum efficiencies (8.4 percent for flat-plate, 19.6 percent for concentrator systems).
2. There is much greater incentive to strive for high cell efficiencies in concentrator systems than in flat-plate systems.
3. Very high manufacturing costs can be tolerated for high efficiency, high concentration-ratio cells.

The numbers in Table 5.1 are of course sensitive to the assumptions made for plant costs, but the qualitative conclusions listed above are not. DeMeo (1978), for example, concluded that future large-scale photovoltaic power plants will likely use either very low-cost flat-plate systems or very high-efficiency concentrator systems. In Chaps. 11 and 12, some exotic and expensive concentrator cells and systems will be described; their development is justified by the preceding economic analysis. One may get some additional feeling for the relative costs of concentrator and flat-plate systems from the following figures. Eight manufacturers made independent projections for the cost of concentrator arrays (concentrators, cells, and tracking systems) in 1976 dollars (Maycock, 1978) for an assumed production rate of 10 MW/year. The projections ranged between $1.59 and $2.31 per peak watt. These prices were several times lower than the flat-plate array costs.

One must not leave this subject without two sobering throughts. First, although concentrator systems can accept very high cell costs, providing cells at costs lower than the acceptable maximum would cause little reduction in the cost of the electricity generated. Second, the unit-area cost of concentrators appears to have less potential for significant reduction through technological innovations than the area cost of solar cells. In the long run, the

Table 5.1 Maximum acceptable cell cost per m²

η			8.4%	10%	15%	19.6%	25%	30%
Flat-plate			$0	$5.5	$23	$40	$59	$76.5
Concentrator		$X = 10$	$0.06/kWh goal cannot be attained with any cell cost if $\eta < 19.6\%$			$0	$187	$365
		$X = 100$				$0	$1870	$3650
		$X = 1000$				$0	$18,700	$36,500

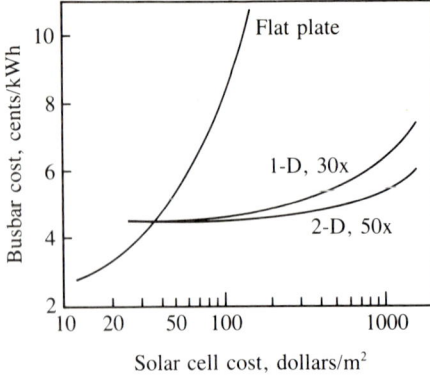

Figure 5.3 Cost of electricity as a function of per-unit area cost of solar cells.

flat-plate systems may yet prove to be the more economical. These conclusions can be easily evaluated using Eq. (5.2.1). Hovel (1978) presented a similar analysis, whose results are shown in Fig. 5.3.

5.3 CONCENTRATOR SOLAR CELLS

The structures and theory of concentrator solar cells are basically no different from those of nonconcentrator cells discussed in Chap. 3. To be sure, there are some subtle differences in the theory, but we shall ignore them for simplicity (Fossum, 1978). Figure 5.4a shows the measured and extrapolated J_{sc}, V_{oc}, and η of a small Si solar cell at constant temperature. J_{sc} is seen to increase in proportion to the concentration ratio, implying a constant collection efficiency. Because of increased J_{sc}, V_{oc} [Eq. (3.7.2)] also increases with increasing concentration. The slope of the V_{oc} curve decreases at high concentrations due to the high-level injection effect and can be improved by using a back-surface field cell (Hu, 1978). Figure 3.16 shows that the fill factor also increases with increasing J_{sc}. Therefore, the conversion efficiency η rises with increasing concentration ratio.

Large area cells typically show a peak in efficiency as seen in Fig. 5.4b. At high concentrations, the cell current is large and a considerable voltage is lost due to the series resistance in the cell (Sec. 3.8). Depending on the value of the cell series resistance, the efficiency would peak at different concentration ratios. By using a closely spaced, fine metal grid for the front contact, RCA (O'Donnell, 1978) has developed a Si solar cell whose efficiency peaks at about 18 percent at 300× concentration. Figure 5.4c shows the calculated efficiency for a GaAs cell with the series resistance as a parameter.

Several novel cell structures to be described in Chap. 11 have exceedingly small series resistance and can operate at up to 1000× concentration.

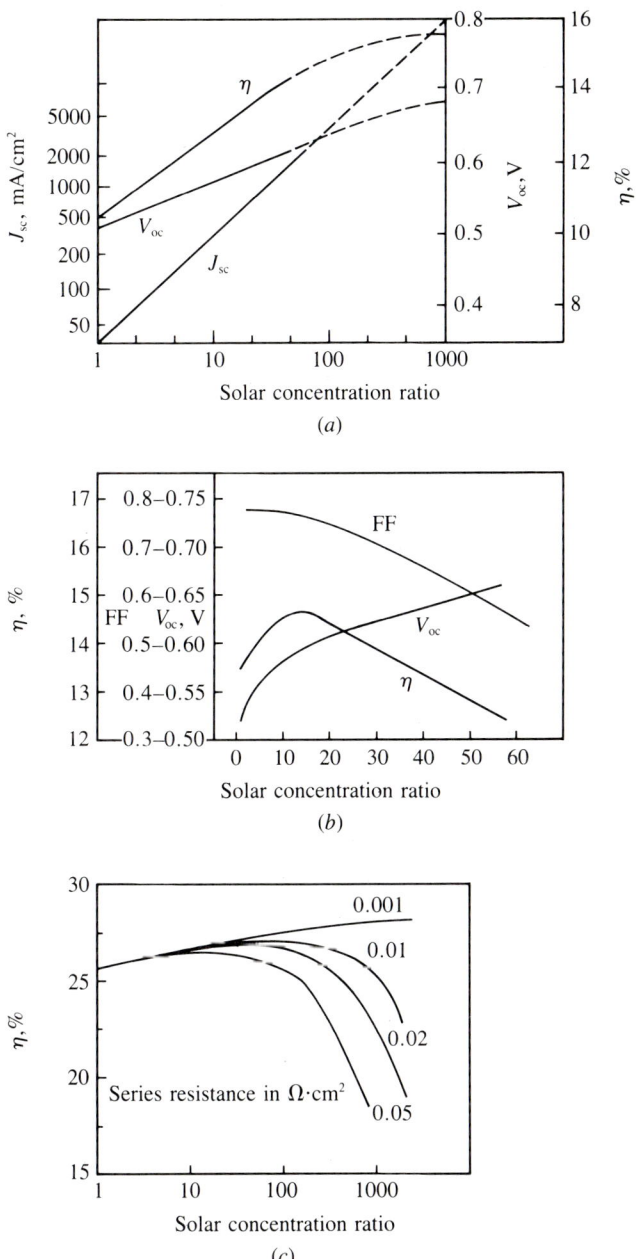

Figure 5.4 Conversion efficiency as a function of concentration ratio. (*a*) A small test solar cell with minimal series resistance; measurements and extrapolations. (*b*) A typical large-area solar cell. (*c*) Calculated efficiency of a GaAs cell.

5.4 COOLING AND COLLECTION OF THERMAL ENERGY

Solar concentration necessarily raises the temperature of the cell. Passive or active cooling must be provided in order to retain the high cell efficiency. This adds to the cost of the system. On the other hand, thermal energy becomes a by-product of concentrator photovoltaic systems. This may increase the economic values of the systems.

The temperature of the cell is determined by the thermal power influx and the thermal resistance of the heat path R_θ by

$$T_c \approx T_a + 1(\text{kW/m}^2) \cdot X \cdot \text{cell area (m}^2) \cdot (1 - \eta) \cdot R_\theta(°\text{C/kW}) \tag{5.4.1}$$

where T_c is the cell temperature, T_a is the temperature of the ambient air or liquid coolant, X is the concentration ratio, and η is the cell efficiency. Increasing X causes T_c to rise. With increasing cell temperature, I_{sc} tends to rise due to improved carrier lifetimes but the fill factor and the open-circuit voltage decrease. The temperature dependence of V_{oc} was discussed in Sec. 3.8. The net effect is that the silicon cells' efficiencies follow

$$\eta(T_c) \approx \eta(25°\text{C}) \cdot [1 - 0.004(T_c - 25)] \tag{5.4.2}$$

Clearly, minimizing T_c would maximize η but would also maximize the heat-exchanger cost.

Edenburn (1980) has made a comprehensive study of the optimization of heat exchanger designs. His basic conclusions are:

1. Passive cooling is effective and more economical than active cooling for 2-D concentrators (point focus) having lens sizes smaller than 0.1 m². Cooling effectiveness depends on the lens size but is quite insensitive to the concentration ratio because the available heat-exchanger-to-air interface area is determined by the lens size. For example, at a wind speed of 3 m/s, the cell temperature can be $T_a + 73°\text{C}$ at $X = 170$ and $T_a + 85°\text{C}$ at $X = 3400$. The cost of the optimal heat exchanger would be about $10 for each m² of lens size in 1980 dollars.
2. Active cooling is preferred for 2-D concentrators with lens areas much larger than 0.1 m² and for 1-D concentrators (line focus). The equipment cost is about $8 per m² of collector area and the energy use is about 7 percent of the cell output.

Edenburn's study suggested that an optimal system may have T_c as high as 80 to 100°C. At this temperature, the thermal energy may be collected and used rather efficiently for space heating or water heating. Where the thermal energy can be used readily, as is the case with the concentrator photovoltaic systems installed on or near residences or other buildings, studies have shown that the total thermal energy produced by the system has about the same value as the electricity (Boes, 1980).

GaAs cells lose their efficiencies with increasing T at about half the rate of Si cells (Sec. 3.8). Therefore they can be operated at higher temperatures. This tends to reduce the heat-exchanger cost and allows GaAs solar cell systems to provide thermal energy at higher temperatures for industrial process heat or for driving thermal engines. For example a $1000\times$ GaAs system employing passive cooling and having 18 percent module efficiency has been reported (Kaminar et al., 1982). In it the junction temperature is 50°C when the ambient temperature is 28°C.

5.5 SUMMARY

Concentrator solar cell systems trade the added costs of the optical concentrators for the costs of solar cells. There are many types of concentrators, but the fresnel lens and the parabolic trough reflector have received the most attention for use in photovoltaic systems. Since a concentrator system uses a much smaller quantity of solar cells than a flat-plate system of the same capacity, relatively more extensive engineering efforts and expensive manufacturing technologies can be applied to improving the efficiencies of concentrator solar cells. Further, a solar cell's efficiency increases with increasing light intensity until the series resistance of the cell causes this trend to be reversed. Many cell designs that minimize the effects of the series resistance are discussed in Chap. 11. Adequate means of cooling must be provided for concentrator solar cells. In some cases, the total thermal energy collected from a concentrator solar cell system has about the same commercial value as the total electricity generated by that system.

REFERENCES

Boes, E. C. (1980), Record, 14th IEEE Photovoltaic Spec. Conf., 994.
Born and Wolf (1970), *Principles of Optics,* Academic Press.
DeMeo, E. A., and Bos, P. B. (1978), Elec. Power Res. Inst. Report ER-589-SR.
Edenburn, M. W. (1980), Record, 14th IEEE Photovoltaic Spec. Conf., 771.
Fossum, J. G., Burgess, E. L., and Lindholm, F. A. (1978), *Solid-State Elec.,* Vol. 21, 729.
Hill, J. M., and Perry, E. H. (1977), Proc. 1977 Annual Meeting American Sec. of Inter'l Solar Energy Society, 37.
Hovel, H. J. (1978), *IBM J. Rev. Develop.,* Vol. 22, 112.
Hu, C., and Drowley, C. (1978), Record, 13th IEEE Photovoltaic Spec. Conf., 786.
James, L. W., and Williams, J. K. (1978), Record, 13th IEEE Photovoltaic Spec. Conf., 673.

Kaminar, N., Borden, P., Gregory, P., Grovner, M., and LaRoue, R. (1982), 16th IEEE Photovoltaic Spec. Conf. (to be published).
Ludman, J. E. (1982), *Applied Optics,* Vol. 21, 3057.
Maycock, D. D. (1978), Record, 13th IEEE Photovoltaic Spec. Conf., 5.
Meinel, A. B., and Meinel M. P. (1976), *Applied Solar Energy,* Addison-Wesley.
O'Donnell, D. T. et al. (1978), Record, 13th IEEE Photovoltaic Spec. Conf., 804.
Rapp, C. F., and Boling, N. L. (1978), Record, 13th IEEE Photovoltaic Spec. Conf., 690.
Swedlow (1976), Report No. 873, Swedlow, Inc., Garden Grove, Calif.
Welford, W. T., and Winston, R. (1978), *The Optics of Nonimaging Concentrators: Light and Solar Energy,* Academic Press.
Winston, R. (1974), *Solar Energy,* Vol. 16, No. 2, 89.

PROBLEMS

5.1 *Compound parabolic (Winston) concentrator* Show with words and a drawing that incident light beams within the acceptance angle of the compound parabolic concentrator reach the solar cell after at most one reflection.

5.3 *Energy cost as function of cell cost* Using Eq. (5.2.1), calculate the cost/kWh as function of the cost per m^2 of cells.

5.3 *Calculation of energy costs* Redo the calculations that result in Table 5.1, using your own estimates for the plant cost and other relevant figures or assume \$20/m^2 for a flat-plate system for land, support structure, etc., and \$100/m^2 for a concentrator system for land, concentrator, tracking structure, etc. Discuss how these changes impact the conclusions indicated in the text following Table 5.1.

5.4 *Effect of series resistance*

(a) Plot the *I-V* characteristics of a typical 10-cm^2 Si solar cell at 200× sun similar to Fig. 3.15a. Show the current and voltage scales.

(b) Draw the equivalent circuit of the cell including some internal series resistance R_s.

(c) In the plot of (a), add two more curves for $R_s = 0.01$ Ω and $R_s = 0.05$. Graphically estimate the cell efficiency at $R_s = 0$, 0.01 Ω, and 0.05 Ω.

5.5 *Passive cooling* Explain with drawings and words why passive cooling can be effective for fresnel-lens concentrator systems regardless of the concentration ratio and why it would *not* be attractive for parabolic-trough concentrator systems.

CHAPTER

SIX

POWER CONDITIONING, ENERGY STORAGE, AND GRID CONNECTION

CHAPTER OUTLINE

6.1 MAXIMUM-POWER-POINT TRACKING
6.2 PRINCIPLES OF MAXIMUM-POWER-POINT TRACKERS
6.3 STAND-ALONE INVERTERS
6.4 INVERTERS FUNCTIONING WITH POWER GRID
6.5 COSTS OF POWER CONDITIONERS
6.6 ISSUES OF ENERGY STORAGE
6.7 ENERGY-STORAGE TECHNOLOGIES
6.8 SUMMARY
REFERENCES
PROBLEMS

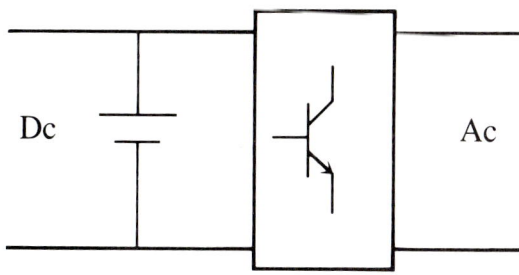

The electricity generated by a solar cell array is often processed electronically before it is fed to the load. The collective name for all the processing that may be performed is *power conditioning*. There may be some simple circuitry for protecting the array and the load from overvoltage and overcurrent. In the case of a battery load, a circuit that prevents overcharging and thus damaging the batteries is often employed. Only two major power-conditioning functions warrant discussion here. They are dc-to-ac inversion, and maximum-power-point tracking.

Dc-to-ac inversion is performed by an *inverter* that transforms the dc power generated by the solar cell array into ac power. This process is obviously necessary if the load, such as a group of electrical appliances, requires ac power. A farther-reaching reason for dc-to-ac conversion is to connect a solar cell array with the utility power grid, either at a central photovoltaic generation station or at a single-family residence with a rooftop solar cell array. Such a connection is highly attractive if the utility power grid is available. It solves two major problems of photovoltaic power generation—a guaranteed supply of power at all times and the storage of excess energy generated.

Whether the load requires ac or dc power, it is desirable to operate the solar cell array at its maximum power point (see Fig. 3.15) at all times. The array *I-V* characteristics change with light intensity and temperature, and the load in general changes unpredictably with time (due to load equipment's being turned on and off). What is desirable is a circuit that electronically and automatically matches the load with the array, in other words a maximum-power-point tracker.

Since photovoltaic power generation is intermittent in nature, some means of energy storage or power back-up is usually necessary. For stand-alone systems such as those located at remote sites, battery storage and/or a back-up diesel generator are commonly employed.

When the utility power is available as the back-up, as is the case in most regions of industrialized countries, an alternative to the expensive battery storage exists. The excess power generated by the solar cell array may be sold to and fed into the utility power grid. Such an integration with the power grid is natural for photovoltaic central generation stations, and it is economically attractive for small, residence-based systems, too. In a sense, the utility grid serves as the storage medium for the photovoltaic power in this scheme.

This brief introduction is summarized in Fig. 6.1. The simplest solar cell systems would involve none of the boxed items. Other systems could involve all of the boxed items.

6.1 MAXIMUM-POWER-POINT TRACKING

A solar cell array consists of many solar cells connected in a series/parallel fashion. The array has *I-V* characteristics similar to those of a single solar cell

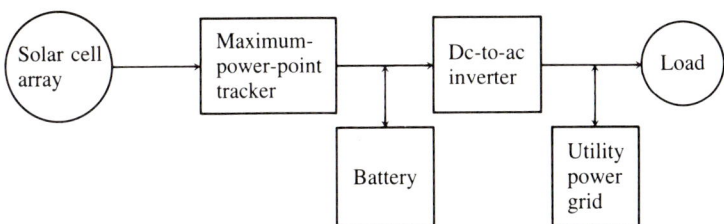

Figure 6.1 Block diagram illustrating power conditioning, energy storage, and integration with power grid. A given system may have all, some, or none of the four boxed items. The arrows denote the directions of power flows.

(Fig. 6.2). The *I-V* characteristics change in response to the variations in insolation. Each *I-V* curve has a maximum-power point (see Sec. 3.7), at which the product of the array voltage and current is maximized. It is clearly advantageous always to operate the array at its maximum-power point, hence the desire for maximum-power-point tracking.

Maximum-power-point tracking is not always employed nor is it always cost-effective. Consider a common type of load, the constant-voltage load such as a battery (or a connection to the utility grid through an inverter, to be discussed in Sec. 6.4). In Fig. 6.2 a battery voltage is chosen such that the array operates at its maximum-power point at noon on a clear day. At other times of the day or when the sky is not clear, the chosen battery voltage would cause the array to deliver less, but not far less, than the maximum possible power. This observation raises the question of whether the increased energy output can justify the cost of a maximum-power-point tracker.

An exact analysis of the benefit of tracking would be difficult because the maximum-power point also varies as a result of temperature changes and,

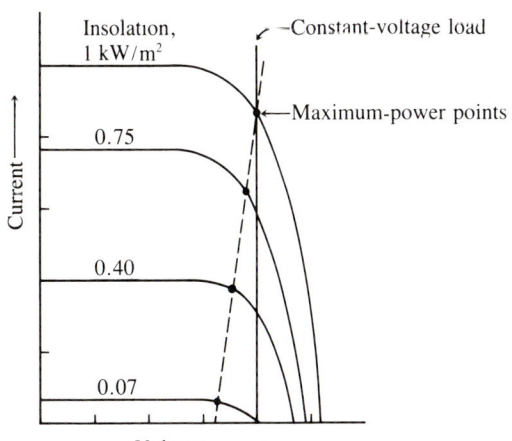

Figure 6.2 Typical *I-V* characteristics of a solar cell array. The maximum-power points vary with insolation as well as with temperature. A constant-voltage load such as a battery cannot extract the maximum power under all conditions.

possibly, aging of the array. Further, a battery is not an ideal constant voltage load. Its terminal voltage can vary by 30 percent depending on the state of charge and the charging or discharging current. Consequently, a maximum-power-point tracker may boost the annual energy output by 5 to 20 percent depending on the overall design and location of the solar cell system and the efficiency of the tracker.

At the present, smaller systems having battery storage do not employ maximum-power-point tracking. Even many of the very large demonstration systems were designed without tracking. In the future, maximum-power-point tracking will likely be attractive for solar cell systems connected to the utility power grid because the inverter should be able to perform the maximum-power-point tracking function at a small additional cost.

6.2 PRINCIPLES OF MAXIMUM-POWER-POINT TRACKERS

Assuming that the maximum-power point is at a lower voltage than the voltage of the battery load, what can be done about the mismatch? Obviously a dc "transformer" is needed. The function of dc voltage transformation is performed by a group of circuits known as *choppers*, or dc-to-dc converters.

Figure 6.3 shows one such circuit. The transistor operates in the *switch-*

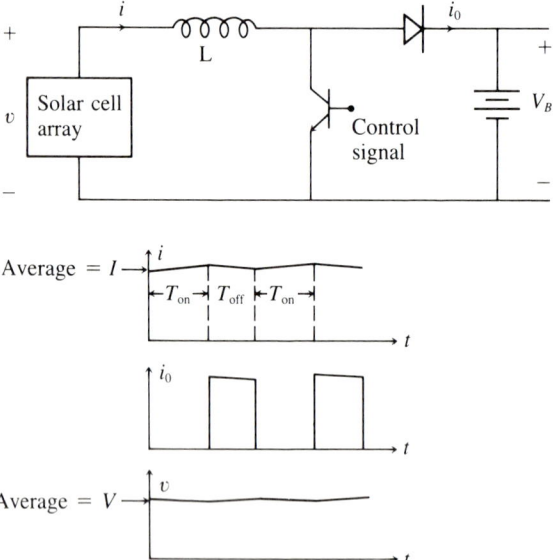

Figure 6.3 A maximum-power-point tracking circuit. By changing the ratio T_{on}/T_{off} the array may operate at any voltage below the battery voltage V_B.

ing mode. It is either off (open-circuit) or operating in the saturation region approximating a short circuit (with at most a few volts of voltage drop). When the transistor is on, the diode is reverse-biased and nonconducting and the array current flows through the inductor and the transistor. During this period, the current increases and so does the energy stored in the inductor, $\frac{1}{2}Li^2$. When the transistor is turned off, the array current flows through the inductor, the diode, and into the battery. During this period the difference between v and V_B (V_B shall be shown to be larger than v) causes a decrease in the inductor current. The goal, of course, is to set the average array current I at the maximum-power point. As i fluctuates around I on the array I-V curve, the array output voltage fluctuates around the maximum-power-point voltage, V.

Over one period, the array output energy must be equal to the energy delivered to the battery. Neglecting power losses in the inductor, transistor, and diode,

$$V \cdot I \cdot (T_{on} + T_{off}) = V_B \cdot I \cdot T_{off} \quad (6.2.1)$$

Canceling I from both sides,

$$V = \frac{T_{off}}{T_{on} + T_{off}} V_B \quad (6.2.2)$$

Therefore, the array can be made to operate at any voltage below the battery voltage V_B simply by adjusting the ratio T_{on}/T_{off}.

How do we determine when V and I have been set at the maximum-power point? Probably the simplest method is to recognize that at the maximum-power point the derivative of power ($v \cdot i$) with respect to v or i is equal to zero.

$$\frac{d(vi)}{dv} = v\frac{di}{dv} + i = 0 \quad (6.2.3)$$

$$\therefore \frac{di}{dv} = -\frac{i}{v} \approx -\frac{I}{V} \quad (6.2.4)$$

Since v and i continually fluctuate (see Fig. 6.3), the derivative di/dv can be evaluated every cycle and compared to I/V. When they are equal, the maximum-power point has been found [Eq. (6.2.4)]. If di/dv is larger than $-I/V$, then V is larger than the maximum-power-point voltage and must be lowered by increasing the ratio T_{on}/T_{off} (Landsman, 1978), and vice versa.

To avoid excessive deviations from the maximum-power point, the fluctuations in i and v must be limited by using a large inductance and/or a small $T_{on} + T_{off}$ period. Transistors are available up to 100 A and 500 V. For higher power capability, one can use thyristors, also known as SCRs (for silicon controlled rectifiers).

There are several chopper circuit configurations (Westinghouse, 1979 and Dewan, 1975). The circuit shown in Fig. 6.3 allows the array to operate

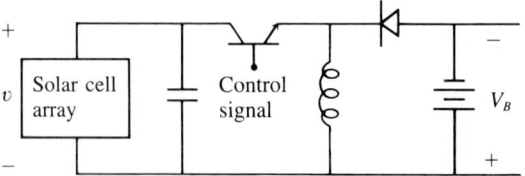

Figure 6.4 A maximum-power-point tracking circuit that allows the array to operate above and below V_B. $V = V_B T_{\text{off}}/T_{\text{on}}$. Note the reversal of the output polarity.

at voltages below V_B. A circuit that allows the array to operate above as well as below V_B is shown in Fig. 6.4. Chopper circuits are manufactured for electric cars and forklift trucks, however the control circuit for tracking the maximum-power point is not readily available yet.

6.3 STAND-ALONE INVERTERS

At a remote village or an energy self-sufficient house not connected to the utility power grid, there may still be a need for a dc-to-ac inverter because electrical appliances in general require ac power. In this case the circuit shown in Fig. 6.5 may be employed.

A clock alternately turns on T_1/T_4 and T_2/T_3 at 50 or 60 Hz frequency. The diodes are needed because the load may be reactive and the load current out of phase with voltage. When v_o is positive (T_1 and T_4 on), i_o may be negative (flowing through the two diodes in parallel with T_1 and T_4).

Although the 180° pulse waveform is simple to generate, it also contains significant amounts of third and fifth harmonics. The 120° pulse waveform contains less harmonics. In particular, it contains no third harmonics. Pulse-width-modulation techniques can further reduce the low-order harmonics.

The circuit can generate three-phase ac power if a third leg of transistors and diodes is added. There are many other types of inverters (Dewan, 1975; Westinghouse, 1979). Such free-standing inverters are the basis of commercial UPS (uninterruptible power supplies), which generate emergency ac power for critical equipment such as computers.

6.4 INVERTERS FUNCTIONING WITH POWER GRID

A utility power grid is an ac busbar of immense magnitude. One grid interconnects all the generation plants and users in a region of a country, in an entire country, or even across national boundaries. The grid is a rather rigid ac voltage source. Tens or hundreds of rotating generators on a grid operate at the same frequency and voltage because each generator is *locked* into the precise speed of rotation and voltage by the frequency and voltage of the grid.

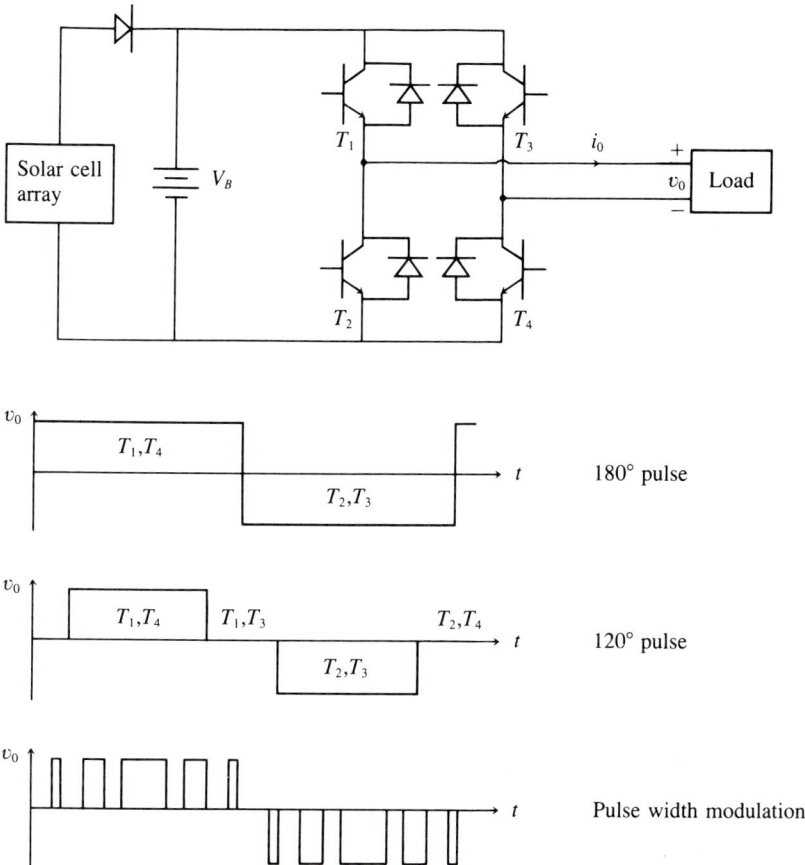

Figure 6.5 A single-phase inverter.

In a similar way, the presence of the power grid simplifies rather than complicates the inverter circuits.

Figure 6.6 shows an inverter that operates into a power grid. Some readers may recognize this circuit as a standard controlled rectifier—the kind widely used for controlling dc motors. Indeed, only the control circuitry of a controlled rectifier needs to be modified before it can be used as an inverter. When operated as an inverter it is called a *synchronous* inverter or a *line-commutated* inverter.

The four identical circuit elements are four thyristors. A thyristor is turned on by applying a short pulse to its gate (corresponding to the base of a transistor). Once turned on it can conduct current in the forward direction like a diode and stays on until the current flowing through it, for any reason,

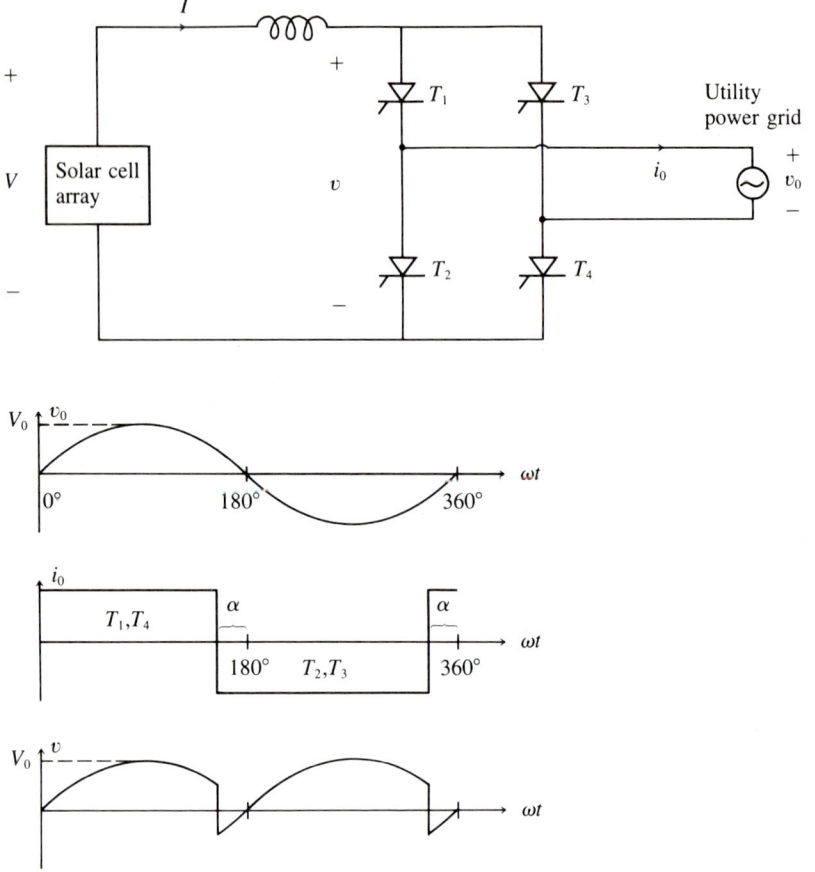

Figure 6.6 An inverter that connects the solar cell array with the utility power grid. The array operating voltage V may be adjusted by changing α.

drops to zero. At that moment it returns to the off state. The thyristors in Fig. 6.6 can only conduct currents in the downward direction.

Assume that the inductance is large enough so that an essentially constant current I flows through the array and the inductor. A control circuit senses the zero crossing of v_o, and turns on T_2 and T_3 $180 - \alpha$ degrees later ($|\alpha| < 90°$). So long as T_2 and T_3 remain on, i_o is negative regardless of the polarity of v_o. At $\omega t = 360° - \alpha$, T_1 and T_4 are turned on. The polarity of v_o is negative at that time, driving a current through the short circuit formed by T_4 and T_2 (also T_3 and T_1), in such directions as to turn T_2 (T_3) off. In the next half cycle, i_o is positive. Since α is less than 90° and probably close to 0°, i_o is nearly in phase with v_o. Thus power is being fed into the grid from the solar cell array.

Turning our attention to the dc side of the circuit in Fig. 6.6, whenever T_1 and T_4 are on, $v = v_o$. Whenever T_2 and T_3 are on, $v = -v_o$. The average value of v is

$$V \equiv \frac{1}{\pi} \int_{-\alpha}^{\pi-\alpha} V_o \sin \omega t \, d(\omega t)$$

$$= \frac{2V_o}{\pi} \cos \alpha \qquad (6.4.1)$$

This is also the average voltage at the solar cell array since there is no average (dc) voltage across an ideal inductor, whose dc impedance is zero. V_o and α, through Eq. (6.4.1), determine the array operating voltage, V. V and the I-V curve of the array determine the array current I. The inverter then presents a (controllable) constant voltage load to the array, much like a battery.

The power-conversion efficiencies of inverters range between 90 and 98 percent, depending on the size of the inverter. Adding a third thyristor string would allow the inverter to connect the array to a three-phase power grid. Equation (6.4.1) still holds, but the i_o waveform becomes an ac pulse train of 120° pulses. From Fig. 6.6, the inverter generates ac power at a power angle of simply α. It has been suggested (Landsman, 1981) that by reducing the inductance and thus allowing the inductor current to vary with time, both the power factor and the current harmonics can be improved. In that case, a large capacitor should be connected in parallel with the solar cell array so that the array current and voltage remain fixed at the maximum-power point essentially free of ripples.

6.5 COSTS OF POWER CONDITIONERS

Both choppers for electric car control (similar to the maximum-power-point trackers) and controlled rectifiers (similar to the synchronous inverters) cost less than $1 per watt in the kilowatt range. Larger units cost less per watt. It should be possible to incorporate maximum-power-point tracking in an inverter at nominal added cost. For maximum-power-point tracking, the array operating voltage must be electronically controllable (see Sec. 6.3). In the case of a synchronous inverter, this can be achieved by controlling α as indicated in Eq. (6.4.1). A control circuit is needed to vary α until the array voltage and current are set at the maximum-power point, i.e., when $dI/dV = -I/V$.

Figure 6.7 shows the photovoltaic power conditioner/inverter costs versus unit power rating as estimated by several manufacturers (Jones, 1978). The actual costs of power-conditioning and control systems of six 16- to 135-kW$_{pk}$ systems completed in 1981 ranged between $0.46 and $2.70 per

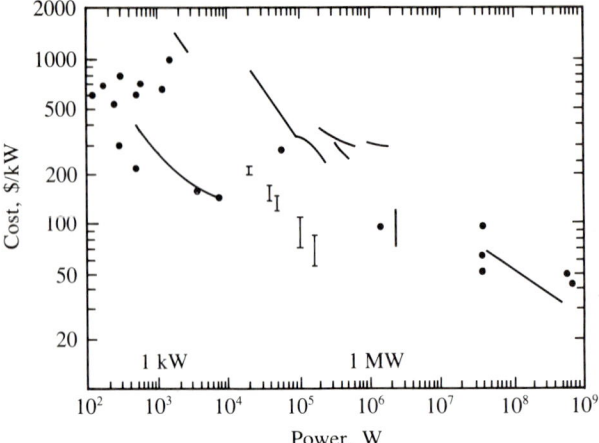

Figure 6.7 Solar cell power conditioner/inverter cost versus unit power rating. Each symbol represents estimates made by a different manufacturer in 1975 dollars. (*Jones, 1978*).

W_{pk}, representing 2.2 and 9 percent of the total system costs, respectively (Burgess, 1981). The costs did not correlate with the power ratings of the systems, although the lowest cost was achieved in the largest system. The cost of power-conditioning equipment is expected to drop, simply because of increased volume of production, and to stabilize around 5 percent of the total system cost in the future.

6.6 ISSUES OF ENERGY STORAGE

Only the storage of *electrical* energy is considered here. Some photovoltaic systems also generate thermal energy, but the storage of solar thermal energy is well covered by other books.

Energy storage is not necessary in some systems, such as those for agricultural water pumping. In general, there is a need for energy storage in a photovoltaic system that is not connected to a utility grid. For these small and medium-sized systems, the lead-acid battery is currently the only practical means of storage. In the future, some advanced battery may be the choice. Less conventional storage elements such as flywheels are also being investigated.

Because of the high cost of batteries, energy storage is sometimes avoided even when the system is separated from the utility power grid. For example, when a photovoltaic system supplements a diesel generator it may be advantageous to select a lower power rating for the photovoltaic system so that even on the sunniest days there is hardly any photovoltaic energy to be stored. This way, no energy storage is needed.

When the solar cell system is connected to a utility grid, it becomes less clear whether to employ energy storage. Energy storage is no longer needed to ensure the uninterrupted supply of electricity. Furthermore, any excess electricity produced by the photovoltaic system can be fed into the utility grid for a credit. Although the buyback credit should be lower than the selling rate charged by the utility, it can be surprisingly high (see Sec. 8.1). Even assuming a very conservative buyback rate of 40 percent of the selling rate, on-site energy storage appears to be uneconomical and unwarranted (Jones, 1980). In fact, an on-site storage system has more value in storing utility-generated energy to avoid the peak-period price under a typical time-of-day rate system than in storing photovoltaic-generated energy to avoid the difference between the buyback and selling rates. Other legal and institutional aspects of utility buyback are discussed in Chap. 8.

Energy storage on the customer/photovoltaic-power-producer's side of the watt-hour meter is not attractive. What about storage on the utility side of the meter? What if the photovoltaic power is generated at utility-run central stations? There, the question becomes essentially divorced from the fact that photovoltaic power is intermittent and fluctuating. Unless photovoltaic power accounts for, say, more than 10 percent of the total generating capacity on a grid—an unlikely situation for some time to come—the fluctuation in photovoltaic output is small compared with the hour-to-hour and day-to-day fluctuations of the electricity demand. The need for energy storage to ensure uninterrupted service is nonexistent or at most questionable. On the other hand, with or without photovoltaic generation by customers or by the utility at central solar cell stations, the use of energy storage to level the load and avoid the high power production costs during peak demand periods (see Fig. 8.4) is beneficial and is being practiced today. A utilities-sponsored study (De Meo, 1978) has also reached the probably obvious conclusion that any energy-storage facilities owned by the utility should be available to the general system rather than dedicated solely to use by photovoltaic generation plants. The presence of large photovoltaic plants could provide added incentive for employing energy storage, however.

Energy-storage technologies suitable for the utilities are conventional pumped hydro, underground pumped hydro, compressed-air storage, and batteries. Pumped hydro storage is the most economical and the only one in large-scale use today. Battery storage is the most attractive new technology. It has the advantage of being modular, a factor that promises a relatively short lead time from order to installation. Other advantages are the capability for adding sections as they are needed, and a wider choice of sites, such as substations. In addition, batteries can respond to momentary fluctuations in load demand and thus improve system regulation and stability. The inverter attached to the battery storage can also generate reactive power, as shown in Fig. 6.6, for power factor correction.

6.7 ENERGY-STORAGE TECHNOLOGIES

For small and intermediate systems, the only practical choice for energy storage today is the lead-acid battery. For the utilities, the most economical storage is pumped hydro storage. Both the small and intermediate systems and the utilities can benefit from advanced batteries. To store half a day's output of a 1-kW_{pk} photovoltaic system requires about 4 kWh of storage. If the storage element is discharged to a depth of 80 percent (in order to lengthen the lifetime of a battery), then 5 kWh of storage should be installed.

Lead-Acid Batteries

When charging current passes through a lead-acid battery, energy is stored by converting the lead sulfate ($PbSO_4$) on the battery electrodes into a mixture of pure lead (Pb), lead dioxide (PbO_2), and sulfuric acid (H_2SO_4). The reaction is reversed when the battery is discharged.

Among the many types of lead-acid batteries available, the type used in forklift trucks and electric cars is most suitable for photovoltaic energy storage. These batteries can last 1500 deep discharge–charge cycles at a reasonable premium in price.

Recent experience with four medium-sized lead-acid battery storage subsystems of photovoltaic systems indicates that the energy efficiency for one discharge–charge cycle is about 80 percent and the subsystem costs range between $82 and $110 per kWh (Brench, 1981). The subsystem costs include installation, shipping, gas detectors, and ventilation equipment. The battery costs are 60 to 70 percent of the total subsystem costs. It has been estimated that automated high-volume production could reduce the cost of suitable lead-acid batteries to $20/kWh in 1975 dollars.

Advanced Batteries

Lead in the lead-acid battery accounts for both the large weight and the high cost of the battery. Advanced batteries avoid the use of expensive materials, as shown in Table 6.1, which lists some of the most intensely pursued developmental batteries.

The sodium-sulfur battery is also known as the *beta* battery because it employs beta-alumina as a solid electrolyte. Molten sodium and sulfur combine to become sodium polysulfide during discharge. The reaction is reversed

Table 6.1 Some advanced batteries

Advanced battery	Active materials cost, $/kWh	Cell energy density, Wh/kg	Demonstrated battery size, kWh
Lead-acid	8.5	25	large
Sodium-sulfur	0.5	150	100
Zinc-chlorine	0.85	100	50
Zinc-bromine	1.7	90	80
Iron redox	1.0	85	20

during charge. This is the only battery in Table 6.1 requiring operation at a high temperature (300 to 350°C). It is also the battery receiving the most research effort. Organizations known to be developing the beta battery include Ford, GE, Dow Chemical, Brown-Boveri, British Rail, Yuasa (Japan), and Compagnie Générale d'Electricité.

The designs and operations of the zinc-chlorine battery and the zinc-bromine battery are similar. During charge zinc is deposited on the negative electrode and bromide ions are oxidized at the positive electrode to become bromine which dissolves in the aqueous electrolyte. A porous separator between the electrodes retards the transport of bromine from the positive to the negative electrode.

The iron redox battery is being developed at GEL Corporation in the United States. During charge iron is deposited on the negative electrode. The SO_4^{-2} ions drift across a membrane to the positive electrode where they combine with the $FeSO_4$ electrolyte to form $Fe_2(SO_4)_3$. The membrane retards the flow of the Fe^{+3} ion.

Other Storage Technologies

Pumped hydro storage is in use for utility load leveling on a grand scale. However, all utility energy storage is of only peripheral relevance to photovoltaic generation, as discussed in Sec. 6.6. To "charge" pumped hydro storage, water is pumped from one lake to another lake located at a higher elevation. When water is discharged into the lower lake through the turbine generator, electricity is generated.

In 1977, there were 57 billion watts and 271 billion kWh of installed pumped hydro storage capacity in the United States. Pumped hydro storage can be developed for about $30/kWh; the United States has already exploited its most attractive sites.

Storage of compressed air in underground caverns for use in Brayton cycle engines and in other ways has been investigated. The only large operating compressed air storage is in Huntorf, Germany. This 290-MW load-leveling system has been in operation since 1978. A 220-MW system in Pike County, Illinois, is scheduled to be completed in 1986 (Lihach, 1982).

A flywheel can store 10 times more energy per unit weight of the system than a lead-acid battery at an estimated cost of $75/kWh. The low weight of the flywheel makes it particularly attractive as an energy source in cars and buses. A prototype photovoltaic flywheel storage system has been developed (Jarvinen, 1981). The flywheel is a 15-inch diameter, 400-pound steel rotor that turns at 7500 rpm to store 1 kWh of energy.

6.8 SUMMARY

For a stand-alone photovoltaic system, the only generally practical technology for energy storage is the lead-acid battery. Because batteries are expensive it is advantageous to use the minimum acceptable amount of storage. The cost-effectiveness of the maximum-power-point tracker is marginal for small systems containing storage batteries.

When a photovoltaic system can be connected to a utility power grid, it is generally more economical to opt for the connection in lieu of battery storage even if the ratio of the energy buyback rate to the selling rate is very low. The dc-to-ac inverter is similar to a controlled rectifier and should be able to perform maximum-power-point tracking as well. The cost of the power-conditioning equipment is and will be around 5 percent of the system cost.

It is uneconomical for the utilities to install storage facilities dedicated to use by centralized or distributed photovoltaic generation plants. The principal reason for storage is load leveling.

REFERENCES

Brench, B. L. (1981), Record, 15th IEEE Photovoltaic Spec. Conf., 642.
De Meo, E. A., and Bos, P. B. (1978), *EPRI ER-589-SR*, Electric Power Research Institute.
Dewan, S. B., and Stranghen, A. (1975), *Power Semiconductor Circuits*, John Wiley & Sons.
Jarvinen, P. O., Brench, B. L., and Rasmusen, N. E. (1981), Record, 15th IEEE Photovoltaic Spec. Conf., 636.
Jones, G. J. (1980), Record, 14th IEEE Photovoltaic Spec. Conf., 1025.
Jones, G. J., and Schueler, D. G. (1978), Record, 13th IEEE Photovoltaic Spec. Conf., 1160.
Landsman, E. E. (1978), Record, 13th IEEE Photovoltaic Spec. Conf., 996.
―――― (1981), Record, 15th IEEE Photovoltaic Spec. Conf., 627.
Lihach, N. (1982), *EPRI Journal*, Oct., 16.
Westinghouse (1979), Semiconductor Division, *Introduction to Power Electronics*.

PROBLEMS

6.1 *Need for storage and inverter* Discuss the needs and considerations for energy storage and dc-to-ac inversion of an agricultural water pumping system, a system serving a remote village with and without a parallel diesel generator, and a rooftop system connected to the utility grid.

6.2 *Step-down dc chopper* Explain the operation of the circuit shown in Fig. 6.4 and show that $V = V_B T_{\text{off}} / T_{\text{on}}$.

6.3 *Maximum-power point* Graphically verify and convince yourself that Eq. (6.2.4) is the correct test for maximum-power point. Graphically follow the control strategy discussed below Eq. (6.2.4) and show that the operating point converges to the maximum-power point.

6.4 *Three-phase inverter* Draw a three-phase (three ac lines) version of the inverter shown in Fig. 6.6. Find the new waveforms.

6.5 *Energy storage for utility plants* Suppose that the cost of power production is the same at all production levels so that there is no incentive for load leveling whatsoever. Discuss whether the utility should then install storage for dedicated use by the photovoltaic generation plants.

CHAPTER
SEVEN
CHARACTERISTICS OF OPERATING CELLS AND SYSTEMS

CHAPTER OUTLINE

7.1 CHARACTERISTICS OF COMMERCIALLY AVAILABLE CELLS
7.2 TYPES OF APPLICATIONS
7.3 OPERATIONAL PHOTOVOLTAIC SYSTEMS AND DEVICES
7.4 SUMMARY
REFERENCES
PROBLEMS

We have previously described how solar cells work and are manufactured, and explained techniques used for sunlight concentration, energy storage, and power conditioning. Here we bring these elements together and discuss characteristics of photovoltaic cells that are now commercially available. Present and future applictions of solar cells are considered in detail to show the design decisions made and to introduce the early data on cell life and system reliability.

7.1 CHARACTERISTICS OF COMMERCIALLY AVAILABLE CELLS

The majority of the terrestrial solar cells now available commercially are flat-plate array cells made of single-crystal silicon having minimum stated efficiencies ranging from 12 to 16 percent. Some of these and other commercial cells appear in Figs. 7.1–7.4. Efficiencies of complete modules of these cells are typically only about 10 percent, owing to reflective losses and to inactive areas between cells and where structural members are located. At least two dozen firms supply cells and modules (see list of suppliers, Appendix 8), and the total worldwide sales for 1982 should be about 10 MW_{pk}. Polycrystalline silicon cell modules having around 8.5 percent efficiency are now available from a few suppliers. Several firms sell cadmium sulfide cells. Prices for encapsulated cell modules run typically twice those of the unmounted cells.

The first commercial gallium arsenide concentrator cells are reportedly available; they are based on a design yielding a demonstrated array efficiency of 18 percent. Approximately 14 percent efficient single-crystal silicon concentrator arrays can be obtained commercially. A recent estimate for a large-scale silicon concentrator array installed, including two-dimensional tracking and foundations (Fig. 7.5), is $12/$W_{pk}$ (1980 dollars) for a 470 kW_{pk} system, with an expected price decline to $2/$W_{pk}$ after 15 MW of such units have been produced and sold. Some firms sell entire PV power systems for specialized applications, such as remote radio transmitter-receivers or navigational aids. Others provide engineering design service to tailor systems for particular applications.

Specification sheets supplied by manufacturers vary widely in content but all typically contain an *I-V* plot for cells, often taken under different illumination conditions and at different temperatures and often identifying the maximum power point. Open-circuit voltage, short-circuit current, efficiency, and mechanical characteristics such as size, weight, and mounting information are also specified. An example is shown in Fig. 7.6.

Listed prices for Si cells up to 12.5 cm in diameter in quantities of thousands range from $9/$W_{pk}$ upward. Individual cells may cost two or three

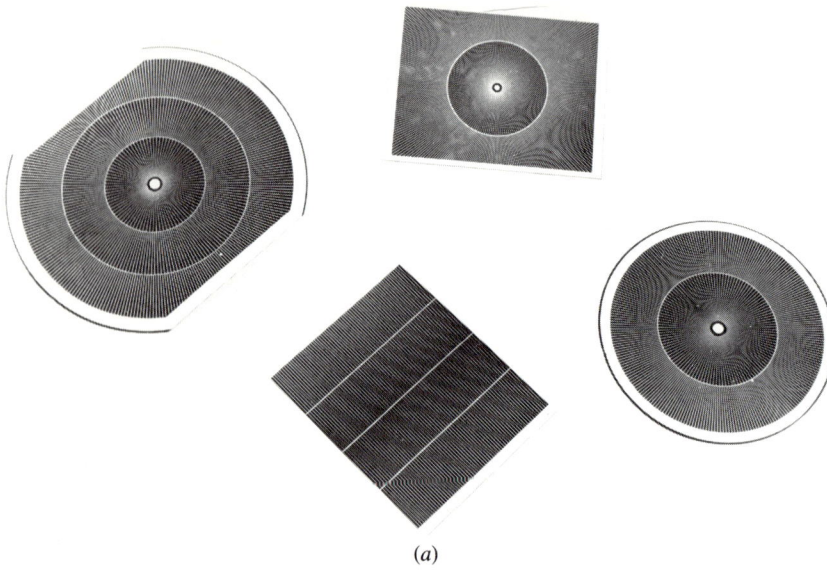

(a)

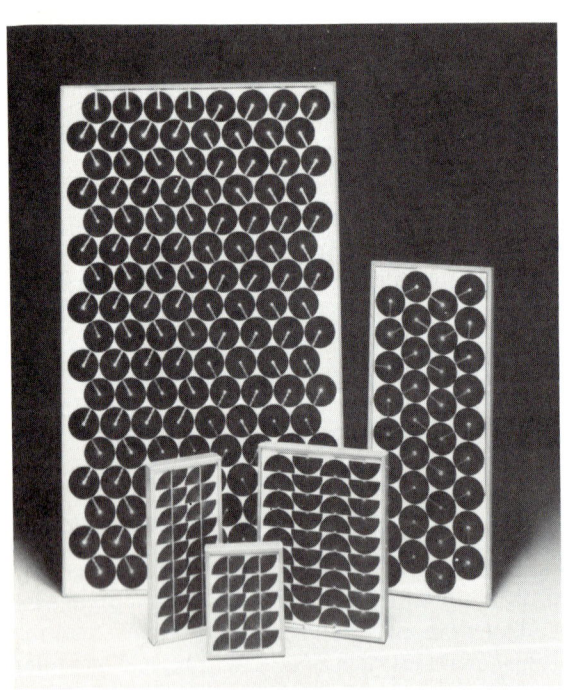

(b)

Figure 7.1 Single-crystal silicon solar cells (a) for one-sun and concentrator use, and one-sun solar cell modules (b). (*Courtesy Applied Solar Energy Corporation, with permission.*)

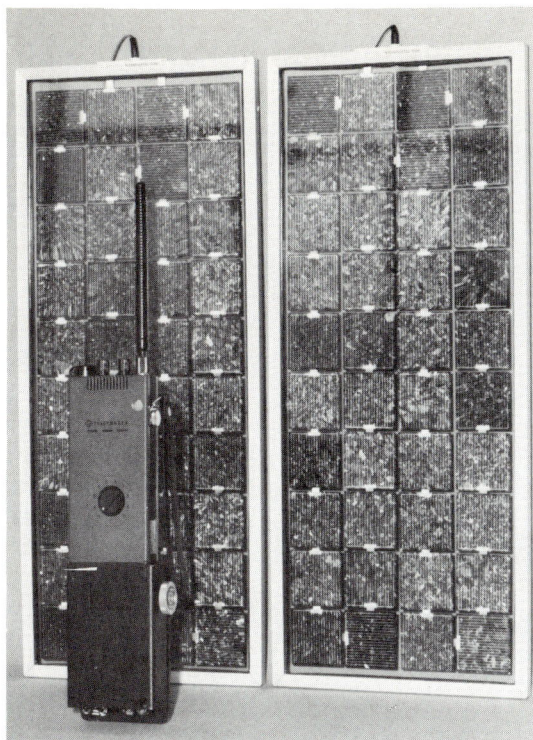

Figure 7.2 Solar modules containing polycrystalline silicon cells for powering communications equipment. (*Courtesy AEG-Telefunken, with permission.*)

Figure 7.3 Panel of eight 24-in^2 CdS/Cu$_x$S cells formed on sheets of float glass coated with conducting tin oxide. The continuous CdS layer has been divided by laser scribing into parallel individual cells approximately $\frac{3}{8}$-in wide. (*Courtesy Photon Power, Inc.*)

132 CHARACTERISTICS OF OPERATING CELLS AND SYSTEMS

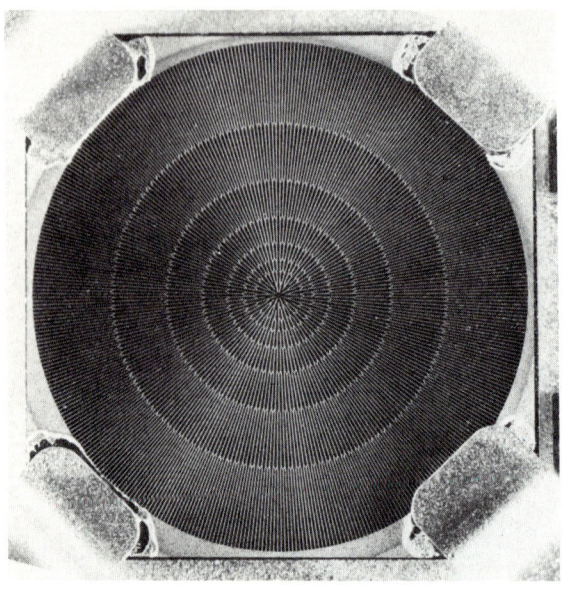

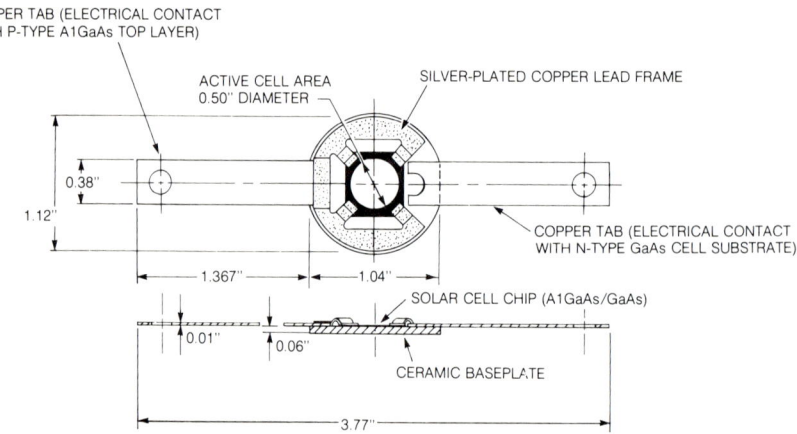

Figure 7.4 Close-up of gallium arsenide concentrator cell and sketch of cell mounting structure that permits heat transfer from cell through ceramic baseplate to passive cooling member. (*From Varian Associates, with permission.*)

times that amount, as with most electronic parts. These prices are appreciably higher than both the current DOE goals and the prices reported for large block purchases by the U. S. government. This disparity likely results from several factors: the somewhat lower costs possible when large markets for cells are assured, the lowered marketing costs associated with bulk sales, and a will-

Figure 7.5 Two-axis tracking concentrator solar cell array. Design intended for use in Soleras and Dallas-Ft. Worth Airport projects. *(Photo by Arco Solar.)*

ingness of some suppliers to buy into the market to establish an early position in the photovoltaic industry.

Demonstrated efficiencies of selected research cells are tabulated in Appendix 5.

7.2 TYPES OF APPLICATIONS

Because of their high cost, terrestrial solar cells have been used to date primarily in governmentally funded test and demonstration projects, or where some unique feature of the PV approach has been important. Examples of the latter are:

• The *portability* of light weight solar cell modules for powering hand-carried communication equipment.

• The *ability to operate far from an electric utility*, necessary in PV systems for cathodic protection of pipelines and bridges or for supplying isolated villages with electricity.

• The *convenience and reduced maintenance* of PV systems exploited in PV-powered marine navigational aids, which formerly required frequent visits for maintenance and delivery of fuel.

In many of these applications, solar cells are already economically competitive with other power sources. In remote areas, power from a solar cell system having battery storage may be more reliable than power transmitted a long distance from an electric utility. If technological innovations and economies of scale in manufacturing bring down the costs, PV-produced power will become increasingly competitive with power from conventional

Electrical Characteristics

	SX-100	SX-110	SX-120
Peak power (Pp)	32	36	40
Voltage at peak power (Vpp)	17	17.25	17.5
Current at peak power (Ipp)	1.9	2.1	2.3
Short-circuit current (Isc)	2.2	2.35	2.5
Open-circuit voltage (Voc)	22	22.25	22.5

NOTES:

1. Panels are measured under full sun illumination (1kW/m^2) at 25°C ±3°C cell temperature. Minimum performance is 2 watts less than peak. The ruling specification is peak watts. For a more detailed explanation, see our *Electrical Performance Measurements* bulletin.

2. Electrical characteristics vary with temperature.

Voltage (Voc)	increases by decreases by	2.4mV/°C/cell	below above	25°C
Current (Isc)	increases by decreases by	25uA/°C/cm^2	above below	25°C
Power (peak)	increases by decreases by	0.4%/°C	below above	25°C

SX-110
Performance at Various Light Intensities, T = 25°C

(curves shown for 1 SUN, 80%, 60%, 40%)

SX-110
Performance at Various Temperatures at AM1 (1kW/m^2)

(curves shown for T = 75°C, T = 50°C, T = 25°C, T = 0°C)

NOTE: These curves are representative of the performance of typical panels at the terminals, without any additional equipment such as diodes, cabling, etc. These curves are intended for reference only. Curves for the SX-100 and SX-120 panels are available from Solarex Marketing.

Mechanical Specifications

Weight:
17.5 lbs (7.9 kg)

Dimensions:
17.5" × 42" × 2.1"
(44.5cm) × (106.7cm) × (5.4cm)

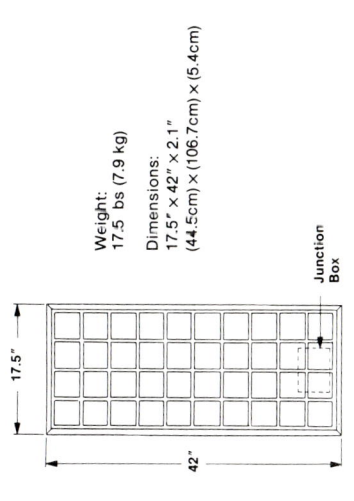

Reliability and Environmental Specifications

These panels are subjected to intensive quality control during manufacture and rigorous testing before shipment. They are designed to meet or exceed the following tests with no performance degradation:

- Repetitive cycling between −40°C and 100°C.
- Prolonged exposure to 90-95% humidity at 70°C.
- Wind loading of over 160 m.p.h.

All SX Series panels are covered by the standard Solarex 5 year limited warranty.

Options and Accessories

Backplates — Anodized aluminum backplate protects the panel in harsh environments. Backplates are available either mounted inside the panel frame at the factory or as components to be mounted onto the panels during field assembly.

Diodes — In-line blocking diode prevents reverse current flow from the panel to the battery during darkness. Bypass diode is available for high voltage systems to provide alternate current path protection.

For multiple panel arrays and large power regulation, contact Solarex Marketing.

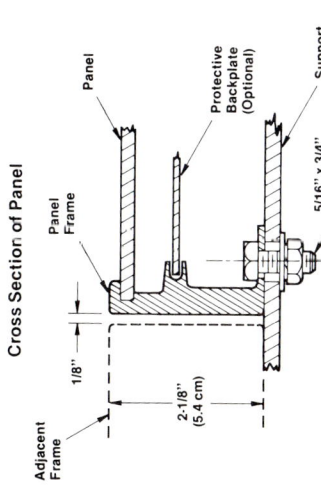

Specifications are subject to change without notice. 6024-1 5/82

Figure 7.6 Sample specification sheet for commercial PV panel containing 40 series-connected 10 cm × 10 cm semicrystalline silicon solar cells. (*Courtesy Solarex Corporation.*)

sources. We will then see the application of solar cells to more common uses, such as residential lighting, refrigeration, and air-conditioning.

Many present types of applications are listed in Table 7.1. As the comments there show, these uses are characterized by quite different combinations of factors—such as whether energy storage is required, whether connection to an electric utility grid is necessary or desirable, and whether one must convert the dc power from the cells to ac power. The diversity of end-uses suggests that different types of PV systems and solar cells will continue to be in demand:

- High-efficiency solar cells will be used where the area for cell deployment is limited or the cost of land is high.
- Cells having only a few percent efficiency may be used in consumer devices such as electronic calculators and watches if their cost is low enough.
- On-site energy storage is not necessary if there is a connection to an electric utility grid that can absorb excess PV power and supply back-up power when PV systems are not functioning. The impacts of such arrangements on utility management and on power costs are substantial.
- Some applications do not require energy storage because the energy demand arises when PV systems are functioning. One example is the residential air-conditioning load, and another is the well-insulated refrigerator whose cooling system can be operated quite intermittently. In some cases, the pumping of irrigation water is necessary only during sunny periods when the PV systems are operating.

Table 7.1 Examples of terrestrial applications of solar cells
Typical power levels for some applications are given in braces

Application	Comments
	I. Large land installations
Building power systems	Provide power to single-family homes, apartment, public, or commercial buildings. Often installed on rooftops. [1 to 100s of kW]
Central power station	Feeds ac power to utility grid for transmission and distribution. [MW to multi-MW]
Forest Service lookout	Supplies lights, communications, refrigeration; energy storage desirable.
Military installations	For remote site; solar and wind power complementarity useful to provide more continuous supply. Grid connection, ac or dc output possibly useful. Storage likely necessary.
Village power system	Water pumping, refrigeration, lighting, communications. [0.1 to 100s of kW]

7.2 TYPES OF APPLICATIONS

I. Large land installations (*continued*)	
Water pumping systems	Mobile systems with no storage, to multi-kW fixed systems that may be grid-connected or independent, operate on ac or dc. [power to multi-kW]
Water purification	[multi-kW]

II. Communications equipment	
Emergency or remote telephone	Energy storage desirable or necessary.
Emergency or portable radio	Storage desirable or necessary. [power to 50 W useful]
Microwave telephone repeater station	Storage necessary; dc output usable.
Military radar installation	Storage necessary. [multi-kW]
Radio broadcast station	Storage or grid connection as back-up necessary; dc output usable. DOE demonstration project as described in text. [multi-kW]
Railroad telephone system	For communications between caboose and engineer on train.
Television receiver in village	[50 W usable]

III. Remotely sited equipment of all kinds	
Alarms	Intrusion, smoke, and fire alarms. Energy storage necessary. [watts]
Cathodic protection system	To prevent corrosion of pipelines, bridges, and structures. Energy storage desirable; usually no grid connection.
Desalinization system	Electricity used for pumping; may be tolerant to fluctuating supply of energy.
Electric cattle fence	Dc output usable; energy storage necessary; usually no grid connection.
Highway dust storm and other warning signs	Energy storage necessary. [watts]
Navigation aids for boats, ships, aircraft	Energy storage necessary. [10s of watts]
Offshore or remotely sited land-based equipment	Lighting, refrigeration, communications equipment; storage desirable.
Scientific instrumentation for for field use	Remote recording or telemetry; storage desirable.

IV. Miscellaneous	
Airplane	Ultra-light "Gossamer Penguin" in 1980 flew powered directly by solar cells without battery storage.
Electronic calculator	Powered by room lights. Energy storage not necessary. Low efficiency cells may be used.
Electronic watch	Solar cells used to recharge battery.

City dwellers in industrialized countries are accustomed to thinking of their electricity supply only when it fails (or when the bill arrives in the mail), as the source has been perfectly elastic, stretching to fill any demand placed upon it. With solar cell systems the situation is different, and each proposed application has required the design of a PV supply tailored for that use. As the market for PV-powered devices grows, standardized cell modules and power-supply packages may be expected, possibly built into the equipment by the original manufacturer. A number of suppliers of solar cells now also sell the balance-of-system (BOS) components such as batteries and power-conditioning equipment, and complete PV systems. In the next section we look at many different applications that have been made and studied to date.

7.3 OPERATIONAL PHOTOVOLTAIC SYSTEMS AND DEVICES

Actual photovoltaic power supplies have ranged in power rating from the multi-kilowatt down to the milliwatt level. Tests have been made of both general-purpose systems supplying power for many different uses and of systems that provide electricity for a single purpose such as the pumping of irrigation water. Here we shall consider the operational experiences reported to date generally on the larger systems, and then examine in more detail a few selected PV supplies at all power levels. A tabulation of many stand-alone, residential, and intermediate load center PV systems is given in Appendix 9.

Large PV Power Systems

Experience with the major DOE-supported PV power systems, listed in Table 7.2, has been summarized by Macomber et al. (1980), who found that reliability has been generally good, particularly with respect to the solar cells and modules themselves. Systems of the single-application type operating at 120 volts or below have been extremely reliable, since failure rates of components have been very low. As an example, in the Mead, Nebraska, agricultural facility (Romaine, 1979), during 27 months of test only 48 of the 2240 modules in the array (2 percent) failed to deliver power, and unscheduled outages of the entire system were quite rare. In the three DOE applications at 240 V dc, reliability has varied from good to poor, but the higher operating voltage may not be the reason. For example, overall system efficiency of the 97,000-cell Mt. Laguna, California, installation has been consistently around 85 percent but some silicon cell cracking (due to overheating and hail damage) and encapsulation delamination has been observed to reduce array output and raise concern over the lifetime of the array, suggesting also the need for

Table 7.2 Some of the higher-power-level DOE-sponsored PV power systems

Peak power	Location	Comments
350 kW	Saudi Arabia	Soleras village project. Complete power supply for villages utilizing tracking concentrator arrays. Operational.
240 kW	Blytheville, Arkansas	Mississippi County Community College (MCCC). System on newly constructed campus employing 40× parabolic concentrating trough, tracked east-west about north-south axis, with 440-V ac inverter-produced output. Excess power to be sold back to utility and with 130°F coolant stream from array to heat campus buildings. Operational.
200 kW	Senatobia, Mississippi	Northwest Mississippi Junior College. Supply for retrofitted campus buildings. Operational.
100 kW	Natural Bridges National Monument, Utah	System providing power at visitor center in remote scenic area for staff residences, maintenance shops, and water sanitation system. Operational June, 1980.
60 kW	Mt. Laguna, California	System providing on average 10% of power used in Air Force radar installation, with approximately 10% diesel-powered generator back-up. Operational August, 1979.
25 kW	Mead, Nebraska	Agricultural PV power system for pumping irrigation water, running fans in grain drying bins, and powering nitrogen fertilizer plant (electric arc discharge in air process). Operational 1977.
15 kW	Bryan, Ohio	DC-only system supplying a daytime commercial radio broadcast station (WBNO) that also has connection to utility grid. Operational August, 1979.
7.3 kW	Concord, Massachusetts	System Test Facility (STF) to provide power, initially at 8 kW level, for typical residential load profile. Operational 1979.
3.5 kW	Schuchuli, Arizona	DC village power supply system for pumping water, lighting, refrigeration, and washing and sewing machines. Operational December, 1978.
1.8 kW	Tangaye, Upper Volta	Village power supply for potable water and grain milling. Operational March, 1979.

accelerated life testing of PV system components. No breakdowns or instabilities have occurred in that system, however.

In a surprising number of these tests, the load devices failed—for example, some refrigerators, weather instruments, and a grain grinder—and in a number of cases trouble was caused by balance-of-system components such as voltage regulators and controls, and batteries. It has been found that highest system efficiencies occurred in cases where the load was relatively constant and well matched to the supply, and where inverters to convert the dc array output to ac for use were not required, as in the daytime commercial radio broadcast station in Bryan, Ohio, where dc itself is used. In cases where people could operate the loads according to their own desires, as in village power systems, sometimes the actual electric loads departed enough from the predicted loads to cause the PV system to be unable to continue normal operation, suggesting the need for more flexible control devices. In systems where the load fluctuates with time, as in the residential systems test facility listed in Table 7.2 (Sacco, 1979), the inverter was too lightly loaded at times and so operated inefficiently, reducing the fraction of the array power that was actually delivered to the load. Inefficiencies resulting from components other than the PV modules have not been negligible and, in the case of systems involving both battery storage and inverters, this fact can result in the need for about a 50 percent larger PV array than one might suppose necessary.

For several reasons it would be unfair to draw sweeping economic conclusions from the costs of these experimental systems, which have ranged from the $20/W level upward. These systems have contained monitoring and data-collection devices that would not be used in later production systems, and they have been made mostly of one-of-a-kind components whose costs would drop in large volume production. Further, because of the lack of broad experience with PV systems, the engineering and construction have been done conservatively and quite carefully, and therefore at premium cost. Perhaps the clearest lessons learned from these tests have related to the balance-of-system components: it is necessary to give considerable attention to increasing the efficiencies of those components and reducing their costs. Standardized system and subsystem design should help in the cost reduction. Improving the efficiency of the solar cells themselves will, of course, permit reduced array sizes and consequently reduced related costs.

Other observations include finding unexpected damage due to lightning even though supposedly adequate protection had been provided by overhead grounded masts to establish cones of protection for the arrays beneath them. Using higher-rated fuses and connecting metal-oxide varistors (surge suppressors) between conductors and ground appears to solve the problem of lightning-caused outages. On the nontechnical side, it was observed during the installation of the 1.8-kW village power system in Tangaye, Upper Volta, that the villagers were eager to have the system and to help in its installation and operation. It is encouraging to note that in the grid-connected Mississippi County Community College project, the buy-back rate proposed by the utility for the purchase of excess PV power is, at least for the first few years of operation, the full industrial purchase rate of roughly $0.02/kWh. If such a policy continues and is followed generally, it will have a positive effect on the economics of grid-connected PV systems.

Electric utilities are showing growing interest in PV systems for primary power production. For example, a 1-MW_{pk} PV facility designed, built, owned, and operated by ARCO Solar will sell power to the Southern California Edison Company. The system employs 108 two-axis tracker units, each of which contains 256 flat-plate silicon solar cell modules. The facility is to be operational by the end of 1982, only eight months after announcement of the start of construction.

Next we consider several different PV systems in somewhat more detail, to provide a more concrete view of this new technology.

Village Power-Supply Systems

Several modest-sized village systems are presently in use (Ratajczak and Bifano, 1979). A 3.5-kW_{pk} system is operating in a Native American (Indian) village in the Southwestern United States. In the 2700-inhabitant village of Tangaye, Upper Volta, a 1.8-kW_{pk} PV system generates power to pump water and grind grain. And a part of the flat-plate PV array for a 5.5-kW_{pk} system in an Indonesian village is shown in Fig. 7.7.

A 350-kW_{pk} system has been built as part of SERI's Project Soleras to demonstrate the feasibility of large PV power systems for isolated villages. In this joint United States–Saudi Arabian venture, the ultimate goal is to design and install 1-MW power systems. The first 350-kW_{pk} Soleras system utilizes the two-axis tracked 40× silicon cell concentrator arrays shown in Fig. 7.5, which follow the sun under computer and sensor control. A single 2.7 × 12 meter array as shown is expected to provide 2.2 kW_{pk} (19.8 A at 114 V) under 0.8 kW/m^2 insolation at 40°C with less than 1 m/s windspeed. In case of a sandstorm or hailstorm, the array will protect its acrylic lenses by turning over to its stowed position with the aluminum heat radiating and convecting structures facing the sky. The entire assembly is environmentally sealed, and sand slides off the lens surfaces upon stowing each day. Measured array efficiency is 10 percent (sunlight to dc power). A second-generation model of this concentrator had 14 percent efficiency when tested at Albuquerque, N. M.

The first of these village systems is described in some detail here to show differences in design philosophies of a stand-alone photovoltaic and a grid-connected electrical system. For example, dc operation was chosen to avoid dc/ac inverter losses, and the PV array and storage batteries were located next

Figure 7.7 Solar cell power supply located in Indonesia. (*Courtesy AEG-Telefunken, with permission.*)

to the water well and its pump, which is the heaviest load in the system. Priorities on types of appliances, hours of use, and the schedule for possible load shedding were decided by the villagers, and some system components were redesigned to increase their efficiency.

Photovoltaic Power System at Schuchuli, Arizona

A 3.5-kW$_{pk}$ photovoltaic power system has been operational since December 1978 in the 15-family, 95-person village of Schuchuli ("Gunsight"), Arizona, on the Papago Indian Reservation (Bifano et al., 1978; Ratajczak et al., 1979). The 120-volt dc system powers the following devices, in the order of priority determined by the villagers:

1. *Village water pump.* A 120-V dc, 2-hp, permanent-magnet jack pump delivering approximately 4.2 m^3/h (1100 gal/h) and connected to a 42-m^3 (11,000-gal) storage tank replaces a diesel-powered pump. Consumption for villagers and livestock varies from 8.7 m^3/day (2300 gal/day) in winter to 19 m^3 (5000 gal/day) in summer, plus about 3.6 m^3/day (960 gal/day) for the single community clothes washer. Pumping time is 3.1 h/day in winter and 5.4 h/day in summer. A control system limits pumping to mid-day hours except in emergency situations.
2. *Lights.* Forty-four 20-watt 120-V dc fluorescent lights were installed with specifically designed 120-V dc/23-kHz inverter ballasts that permit the lamps to produce the same light output as with 120-V ac/60-Hz ballasts. Lights are in houses (two each, one in the kitchen and one in another room), feast house (six), domestic services building (two), church (two), and in the electrical equipment building (four). Previously kerosene lanterns were used.
3. *Refrigerators.* Fifteen personal, 0.11-m^3 (1-ft^3) lockable refrigerators are located together and powered in groups of three by single compressors having $\frac{1}{8}$-hp 120-V dc permanent-magnet motors. Custom-designed, these refrigerators are particularly well insulated with polyurethane foam and their evaporator cold walls contain a gel that is frozen by the refrigerant circulating in the system.
4. *Clothes washer.* One wringer type, having a $\frac{1}{4}$-hp permanent-magnet, 120-V dc motor, is operated from a clock timer up to 12 h/day, 7 days/week.
5. *Sewing machine.* One is available for community use, having a $\frac{1}{8}$-hp 120-V dc universal motor and small light.

The solar cell array consists of twenty-four 1.2 m × 2.4 m flat-plate silicon solar cell panels, arranged in three rows of eight panels each in a 21 m × 30 m fenced and locked area. Tilt angle is adjusted manually four times a year to 3.5° (summer), 26° (spring and fall), and 48° (winter). Panel frame and support structures are designed to withstand 45 m/s (100 mi/h) winds. The 8064 cells are approximately 8.9 cm across and about 15.5 percent efficient. Voltage regulation is achieved by 24 relays that switch strings of PV cells, protected by blocking diodes, to the main bus to keep the voltage constant. System over-voltage and under-voltage conditions are also sensed and cause either the PV array or the loads to be disconnected, and an alarm indicator to light.

Battery storage is provided by fifty-one 2380-A·hr capacity series-connected cells, in parallel with four pilot cells for load management (three having 310-A·h capacities and one having 1055-A·h capacity). Capacities are measured at 77°F and a 500-h discharge rate. The batteries have lead-calcium plates capable of deep-discharge cycle operation. As battery capacity decreases, loads are sequentially shed: At 50 percent depth-of-discharge, the washing and sewing machines are disconnected; at 60 percent, the lights; at 70 percent, the water pump; finally, at 80 percent, the refrigerators. Upon recharging, loads are reacquired in order. The NASA Lewis Research Center computer program for solar cell system sizing, used to design this village system, indicates that with a 20 percent degradation of cell output due to potential darkening of the silicone encapsulant and to dirt accumulation, and a ±20 percent variation from average values of insolation, the maximum depth of discharge of the batteries should be about 60 percent.

Operation of the system is being monitored by automatic recording equipment, and the effects upon the villagers' lives of introducing the PV system are being studied. Cost projections, assuming a moderate growth of market demand for such systems, indicate they can provide power at $1.76/kWh in 1978 dollars on an annualized basis (20-year life, 8 percent discount factor, 5 percent per year escalation of electricity costs over normal inflation). This rate is midway between the $1.55/kWh for 6255 kWh/yr estimated for electricity from the Papago tribe's utility authority and the $1.91/kWh from the nearest private power company. No adverse environmental effects of the system have been identified.

PV-Powered Agricultural Facility at Mead, Nebraska

The 25-kW_{pk} experimental facility at Mead, Nebraska (Hopkinson, 1980), is designed to operate a 20-hp motor 12 h/day to pump 5.7 m^3/min (1500 gal/min) of irrigation water for a 0.32-km^2 (80-acre) field of corn. After the harvest, the array powers two 5-hp fans that force air through two 210-m^3 (6000-bushel) grain-drying bins in which the corn crop is stored. Operation of a 3-kW electric arc discharge-in-air nitrogen fertilizer generator is also scheduled. The facility is composed of two large arrays of solar panels (described below), lead-acid storage batteries having a 90-kWh total capacity, an inverter for changing dc array output to three-phase ac, a dummy electric load for power management and array testing, and control circuitry. The system is tied to the three-phase, 240-V ac utility grid that delivers power or distributes excess power generated by the PV system to other utility customers.

The array consists of two parallel, 113-m-long rows of solar panels located one above the other on a sloping berm forming the edge of the cornfield. At the base of the berm is a 2500-m^3 (2-acre-ft) reservoir into which a pump driven by power from the utility at night deposits well water which is pumped during the day by power from the array and from storage. Supports for the frames holding the PV panels are reinforced concrete piers 76 cm in diameter extending 2.9 m into the soil. The modules are made up of single-crystal silicon cells purchased from two different manufacturers. In each array the cells are connected in series-parallel arrangements to provide outputs at around 150 V dc. In one of the two slightly different arrays, each group of 44 cells is connected in series to form a module having a nominal 16.5-V output. Modules are paralleled in groups of four, making what are termed *quads*, each having a protective diode in parallel across it to carry the current in case a cell becomes open-circuited. Nine of these quads connected in series form a so-called substring having a 150-V output. Finally, three of these substrings are paralleled so their currents add, giving an output of at least 5.9 A at 150 V, or 885 W minimum. Series-protective diodes in the positive legs of the three substrings are used to guard against excessive currents.

Figure 7.8 shows a block diagram of the system. The array, at upper left, feeds the dc bus that can also be connected via relays to the storage battery, an inverter, a 20-hp dc motor coupled to the irrigation water pump, and a bank of resistors serving as a power dump used primarily for test and measurement purposes. The battery charger can be driven from the ac grid when it is necessary to use utility power to charge batteries. The batteries are not intended to carry the system through periods of cloudy weather, but rather to

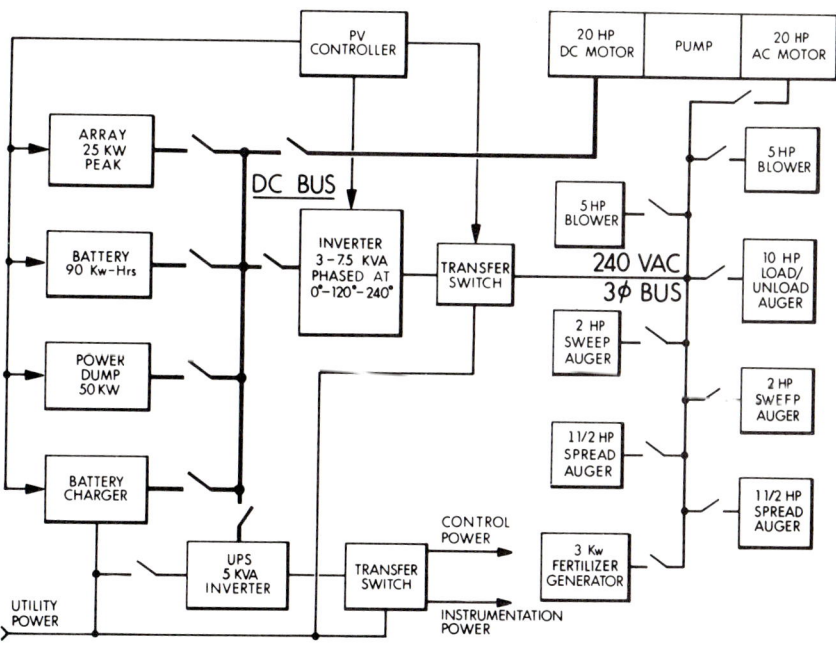

Figure 7.8 Block diagram of Mead, Nebraska, PV-powered agricultural complex. (*Reprinted with permission from Project Rept. C00-4094-10, "The Mead, Nebraska, 25-kW Photovoltaic Power System," W. R. Romaine, Lincoln Lab., M.I.T., January 5, 1979, p. 23, DDC AD-C00-4094-10.*)

square up the curve of output versus time of day: the battery is charged in the middle of the day when the PV array provides excess power, and the battery returns energy during morning and afternoon hours when the insolation is less. Excess PV power can also be fed back into the utility grid after it is converted to three-phase 240-V ac power by the inverter. An uninterruptible power supply (UPS) was included to ensure a stable supply for the control and instrumentation circuitry after it was found that the utility power at this site was subject to frequent brief interruptions.

The electric loads include either an ac or dc motor for the irrigation pump. It was found that the efficiency of the dc motor equalled that of the ac motor, 87 percent, and so the 10 percent power loss in the inverter could be avoided. The 3-kW nitrogen fertilizer generator employs an electric arc in air which produces oxides of nitrogen that are trapped in water containing limestone. The result is water containing calcium nitrate, a desirable form of fertilizer.

Microirrigation Systems

At the other end of the scale of pumping capacities is the portable microirrigation system shown in use in Fig. 7.9. This prototype unit consists of a 240-W flat-panel PV array mounted on a wheelbarrow frame and connected to a submersible pump for raising water from shallow wells.

The units were designed after study showed (Smith, 1977) that there was need for them in the broad, alluvial valleys and deltas of great rivers such as the Nile and the Ganges, where millions of farmers live on very small plots of prime crop land with plentiful water that must, however, be raised to the levels of the fields. The power required for this purpose is surprisingly small: for a one-hectare plot (2.5 acres) planted in a nonrice cereal grain, to lift surface water 1.5 m requires just 320 Wh/day, or about 70 W_{pk}, if one assumes a 50 percent pumping efficiency (Smith and Allison, 1978). Having additional irrigation water would increase yield or permit a second crop, and the added value to the farmer would be greater than the cost of these microirrigation sets even at the 1978 prices prevailing for silicon solar cells made conventionally but in large quantities. A unit price of $1200 is projected when these units are manufactured in quantities of more than 10,000.

The use of storage batteries was avoided because of the 50 lb or so they would add to the weight, the intolerance of lead-acid batteries to deep discharge or the higher cost of discharge-tolerant batteries, the need for the occasional addition of relatively pure water, and the 75 percent turnaround efficiency typically found. The unit shown combined a small vertical turbine pump with a $\frac{1}{3}$-hp permanent-magnet dc motor whose efficiency is 85 percent. A dc motor was chosen because of the relatively low efficiency of fractional-horsepower ac motors, and secondarily because then there is no need for an inverter. A solid-state controller adjusts the load impedance seen by the array to maximize power output as the pump is used at different depths or as the

Figure 7.9 Microirrigation system raising water from a shallow well to irrigate a field. The submersible pump is beneath the spherical floats. *(Courtesy Solar International.)*

insolation changes. Under AM1 conditions, the model shown will lift 0.17 m^3/min (45 gal/min) about 2.4 m, and 0.10 m^3/min (27 gal/min) can be lifted 4.9 m.

These systems could be manufactured in Third World countries. Purchase of the units by the small farmers for whom they were designed would presumably require an initial subsidy of some kind. It has been suggested that a progressive pricing scheme might be instituted, since the farmer who purchased a second unit would be able to pay a higher price more reflective of the true cost of the unit, because of the additional output then being obtained from the farm.

Remote Applications Without Grid Connection

Largest of the growing number of these remote applications is the 100-kW$_{pk}$ system dedicated in June 1980 in Natural Bridges National Monument in southeastern Utah, where the system powers six staff residences, maintenance facilities, a visitors' center, and a water-sanitation system at the 31-km^2 (7600-acre) park, which is the site of three of the world's largest natural sandstone bridges (Jarvinen, 1978). In a 5700-m^2 cleared area, the array field is composed of 48 subfields producing just over 2 kW each. Total module area is 1700 m^2, with 990 m^2 of silicon cell area. Actual conversion efficiency based on gross module area is from 5 to 7 percent. With its battery storage, the back-up of a 40-kW diesel powered generator is expected to be required only 5 to 10 percent of the time, during extended periods of inclement weather.

Other remote applications in use or planned include the following, which are of particular interest.

In a reverse-osmosis water-desalinization system, a 250-W PV array would supply pumps to bring in water to be treated and to maintain the pressure on the water being forced through a semipermeable membrane that blocks passage of the salt. The plant will produce 180 liters of water per hour from 360 liters of well water having a maximum salt content of 1 percent (Telefunken, 1980).

Communications systems employing from ten to several hundred watts of PV-supplied power are in use by park personnel, law enforcement agencies, and commercial firms. A fairly typical installation in the mountains of California is shown in Fig. 7.10. In Lasel, West Germany, an area that formerly

Figure 7.10 Solar cell panel on Martis Peak, California, used to power remote forest lookout communication system.

7.3 OPERATIONAL PHOTOVOLTAIC SYSTEMS AND DEVICES 149

Figure 7.11 Solar-cell-powered TV repeater. *(Courtesy AEG-Telefunken, with permission.)*

Figure 7.12 National Park Service radio communications repeater atop 3186-m-high (10,453-ft) volcanic Lassen Peak in California. Before installation of the solar panel, a helicopter had to deliver charged storage batteries to the site every two months.

had poor television reception because of line-of-sight obstructions is now receiving broadcasts relayed by a PV-powered fill-in transmitter that receives the faint signal from the main transmitter, amplifies it, and rebroadcasts it locally (Fig. 7.11). Radiated power is 2 W and the PV supply is rated at 240 W. To reduce battery requirements, a supplementary wind-driven generator is also included in this system (Telefunken, 1980).

Photovoltaic supplies of from 35 to 150 W are used to power remotely located television receivers for communication and education. Radio and radio-telephone repeater stations in many places around the world are being fitted with PV power supplies, typically at the hundreds-of-watts level. Figure 7.12 shows one such repeater station.

Photovoltaic supplies can, of course, be used in conjunction with solar thermal collector systems to pump the heated water from the collectors to the point of use. Several installations have been made of "total energy systems" like that sketched in Fig. 7.13. This is a simple 5× solar cell concentrator system in an enclosure having a transparent front cover. Air forced through the enclosure is heated and both the electrical output from the cells and the

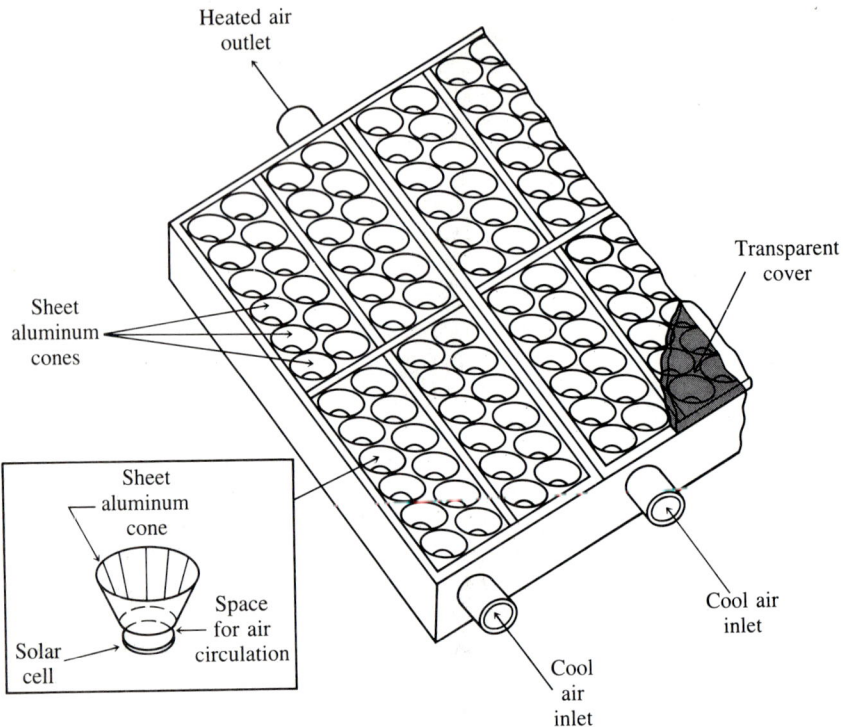

Figure 7.13 Total energy system employing solar cells with some concentration produced by the sheet aluminum cones. Air circulated through the covered boxes can be used to heat a dwelling or commercial building. A single concentrating cone and solar cell are enlarged for clarity. (Drawing is simplified sketch of commercial unit made by Solectro-Thermo, Inc.)

thermal energy of the heated air are available for use. Maximum power outputs for the 1.2×1.2 m^2 (4×4 ft^2) modules are 58 W electrical and 720 W thermal. The heated air is used for space heating in some applications and to preheat water in other installations.

The module just described is dimensioned to permit easy integration into buildings which, in the United States, often employ structural members measuring 4 ft × 8 ft. Another solar cell module that can be installed quite simply is the General Electric photovoltaic shingle, consisting of sealed hexagonal units that can be fastened onto a roof to seal it and produce power.

A final example of a simple application that is not so remote is the 8-W PV supply for a traffic counter located beside some city streets in Glendale, Arizona. Though grid power is available nearby, it is not used since battery storage is included in the system and the substantial cost of making a weatherproof connection to the electricity grid wires is avoided.

Portable Applications

To distinguish "portable" from "movable" someone has said that when a piece of equipment is portable a person can carry it *and* something else at the same time. By this definition, some significant movable PV devices include PV-powered navigation buoys and solar cell power supplies for pleasure boats, both being rated in the 50- to 100-W range. The really portable applications are ones where the use of PV panels provides the user with great mobility by eliminating the need for carrying bulky batteries. A clear example of this use is the field radio system shown in Fig. 7.14, powered by four lightweight panels capable of producing about 40 W_{pk}.

Solar cell–powered watches and even a sun-recharged flashlight are no longer rarities. The PV-powered pocket calculator of Fig. 7.15 can be operated either in sunlight or under artificial indoor lighting, and it needs no batteries because of its intentionally intermittent use. Failure of the only moving part of these devices—the on-off switch—is avoided: no switch is used as there are no batteries to run down. The power consumed by integrated circuits in a calculator and by a liquid-crystal display is so low that less than a milliwatt is adequate, permitting the use of cells having relatively low efficiency. An example is amorphous silicon cells, discussed in Chap. 10, which can be deposited directly on a transparent cover for this application.

Figure 7.14 Portable solar panel powering communications radio transceiver at remote field location. *(Courtesy AEG-Telefunken, with permission.)*

152 CHARACTERISTICS OF OPERATING CELLS AND SYSTEMS

Figure 7.15 Calculator powered in ambient light by amorphous silicon solar cell visible below the liquid-crystal display.

7.4 SUMMARY

Portability, reliability, and operation far from conventional power sources are considerations that have led to many present terrestrial uses of solar cell systems. In large government-supported demonstration projects, the cells have generally performed well, though some problems have been encountered in equipment used with the cells. Most systems used to date have been designed for their specific applications, but increasing use of standardized modules is expected, for economic reasons.

REFERENCES

Bifano, B. J., et al. (1979), Social and Economic Impact of Solar Electricity at Schuchuli Village: A Status Report, DOE/NASA/20485-79/3 (NASA TM-79/94), June.

Bifano, B. J., Ratajczak, A. F., and Nice, A. W. (1978), Design and Fabrication of a Photovoltaic Power System for the Papago Indian Village of Schuchuli, Arizona, DOE/NASA/1022-78/39 (NASA TM-78948), June.

Hopkinson, R. F. (1980), Performance of a PV powered 20-hp dc/ac irrigation system and a 3 kW nitrogen generator, Record, 14th IEEE Photovoltaic Spec. Conf., 115–120.

Jarvinen, P. O. (1978), Natural Bridges National Monument, Utah—Solar Photovoltaic Power System Design, 2nd National Conf. on Technology for Energy Conservation, Albuquerque, N.M., 24–27, January.

Macomber, H. L., et al. (1980), Photovoltaic applications—Past and Future, Record, 14th IEEE Photovoltaic Spec. Conf., 1004–1017.

Ratajczak, A. F., and Bifano, W. J. (1979), Descriptions of Photovoltaic Village Power Systems in the United States and Africa, DOE/NASA/2-485-79/1, April.

Romaine, W. R. (1979), The Mead, Nebraska, 25 kW Photovoltaic Power System, MIT Lincoln Laboratory, Lexington, Mass., C00-4094-10, 5 January.

Sacco, S. B. (1979), Description of the MIT/Lincoln Laboratory Photovoltaic Systems Test Facility, MIT Lincoln Laboratory, Lexington, Mass., C00-4094-41, 30 June.

Smith, D. V. (1977), Photovoltaic Power in Less Developed Countries, MIT Lincoln Laboratory, Lexington, Mass., C00-4094-1, 24 March.

Smith, D. V., and Allison, S. V. (1978), Micro Irrigation with Photovoltaics, MIT Energy Laboratory, Cambridge, Mass., MIT-EL-78-006, April.

Telefunken (1980), Data from technical bulletins supplied by the Telefunken Company.

PROBLEMS

7.1 *Personal electricity use* Table 8.2 in the next chapter lists typical power ratings for electric household appliances. Considering those that you use personally, estimate your electrical energy usage. Compare this with what you actually use, as given on your utility bills, and consider reasons for any difference. Compare these figures with the per capita electric energy usage in the United States (obtain this from Fig. 1.3, assuming the U.S. population is 220 million) and explain the sources of the large discrepancy.

7.2 *PV system ownership and maintenance* Suppose a PV system at your dwelling or business can supply all the electrical energy you use there. Would you want to own and maintain the PV system, or pay more for your electricity to have another party take care of ownership and maintenance matters? If the latter, what percentage increase in the cost of electricity would you be willing to pay for this convenience? Independent of the ownership and maintenance issue, would you want to maintain a connection to the electric utility in order to obtain backup power?

7.3 *Concentrator or flat plate* Discuss the tradeoffs between concentrator and flat-plate PV systems in a variety of applications and locations differing in clarity of sunlight, availability of maintenance personnel, value of a thermal output in addition to electrical output, etc., assuming the concentrator system is 50 percent more efficient than the flat-plate system.

7.4 *Power versus energy* A *Newsweek* (news magazine) article states that ". . . a 50 square feet solar cell panel on the roof can not even generate enough electricity to power a toaster." Is this true? Is it misleading? What is your response to this?

7.5 *Business/applications opportunities* If someone wants to establish a small business in the general field of solar cells, what would you suggest and why? Discuss the opportunities in manufacturing, sales, installation, services, or overlooked or rapidly growing applications.

CHAPTER

EIGHT
ECONOMICS OF PHOTOVOLTAIC POWER

CHAPTER OUTLINE

8.1 SOME GENERAL RULES
BOX: TEN RULES OF THUMB
8.2 COST ANALYSES FOR PHOTOVOLTAIC POWER
8.3 UTILITY ISSUES IN INDUSTRIALIZED COUNTRIES
8.4 ISSUES IN DEVELOPING COUNTRIES
8.5 SUMMARY
REFERENCES
PROBLEMS

The frequent question, "When will solar cells be practical?" is really many questions, and to frame an answer one must consider the viewpoint of the person asking it. The resident of an industrialized country wants to know when the cost of putting solar cells on the roof overhead will be equal to the amount saved because of reduced payments to the energy utility. The businessperson wants to know when photovoltaics might become a profitable enterprise, and the student wonders whether it will soon be a good field in which to work professionally. The utility executive and the energy planner want to know when PV-produced electricity will account for a significant fraction of the large amounts of power and energy they deal with. And, on the low end of the scale of energy magnitudes, the concern in villages may simply be to know when a 100-W unit may actually be installed to pump water for cattle and humans.

From the strictly technological perspective, solar cells are already practical, and they are continually being improved. But the question being asked is really one of *economic* practicality, and the simplest answer to the question is, "When the price of the electricity from the PV system becomes lower than that for electricity from the alternative sources available for this application." For powering spacecraft, solar cells were economical long ago, even at capital acquisition costs around $50,000/kW$_{pk}$, because they weighed much less than the alternative systems that stored energy in batteries or in chemical fuels. And in some remote applications, terrestrial cells are already practical. Examples in Sec. 7.3 are the Schuchuli village and remote forest communication systems where a grid connection is far too expensive and a diesel generator involves high costs for fuel purchase and delivery, and for maintenance.

8.1 SOME GENERAL RULES

One cannot state when the price of PV-produced electricity will have fallen enough to permit truly widespread application, but a few general principles are clear:

1. *As costs of PV-produced electricity are gradually reduced, economic practicality in a particular geographical region occurs first where the cost of electricity from other sources is highest.* The value of the electricity in a grid is highest at the load, after transmission and distribution from the power-generating station, and so in industrialized countries, economic feasibility will occur first at dispersed load sites such as dwellings and light commercial establishments.
2. *Economic feasibility of PV power is highly dependent upon the future prices of electricity from conventional sources.* These prices reflect, as

noted earlier, fuel prices and the rate of inflation. Computer simulations of economic trends typically make high and low assumptions about the expected rise of the price of electricity from the grid and the rate of inflation. These simulations then yield different dates for practicality, depending on the high and low values assumed.

3. *The time at which economic practicality of PV systems will occur is very dependent upon the characteristics of the individual consumer using the PV system.* The location of the user is important because of the geographical variation of insolation and because the price of power from the grid varies greatly. For example, residential electricity in the Northeastern United States costs two or three times what it does in the Southeast. Does the individual user consider and compare with that of the alternative grid power the *life-cycle cost*—total costs for the 20- or 30-year life of the system—or are only the *first-year costs* considered and compared? Is there a governmental tax credit available as an incentive for installation of solar equipment (amounting, for example, in 1980 to 40 percent in U.S. federal income tax, and 55 percent of costs per function in the State of California, subject to certain limitations and conditions)? If so, the amount of that credit will affect the perceived cost and attractiveness of the solar option. The income-tax bracket of the purchaser is also relevant because interest paid on money borrowed to buy and install the PV system may be deductible from income when figuring the tax to be paid.

Other individual choices having economic implications both upon the consumer and upon the electric utility to which the consumer is connected are the degree to which intermittency of the electricity supply will be tolerated and hence the amount of grid back-up that will be sought, the power buyback policy of the utility, whether time-of-day pricing is used by the utility, and whether the utility or the customer will want to own and maintain the PV system. Finally, if the consumer will be able to use the thermal output from the PV system this will obviously tend to favor economic feasibility.

4. *Very different economic and technological criteria apply in the industrialized and the nonindustrialized countries.* In some parts of the planet it may be desirable economically and socially to engage otherwise unemployed persons to adjust PV panels periodically through the day rather than relying on complicated, possibly unreliable electro-mechanical tracking systems. It may be worthwhile in some countries for people to set up portable PV systems in fallow fields temporarily to generate electricity for local use, storage, or for feeding into a utility grid.

In the remainder of this chapter we will draw from analyses of PV-system prospects at specific locations in the United States, carried out both by governmental agencies and by private firms, to illustrate expectations about

TEN RULES OF THUMB

Here are some *approximate* quantities or relations useful for estimating the characteristics of photovoltaic systems.

1. The maximum peak terrestrial sunlight intensity is 1000 W/m^2, and the average sunlight intensity year-round and worldwide is approximately one-fifth of the peak intensity, or about 200 W/m^2. See Chap. 2 and the illustration below.

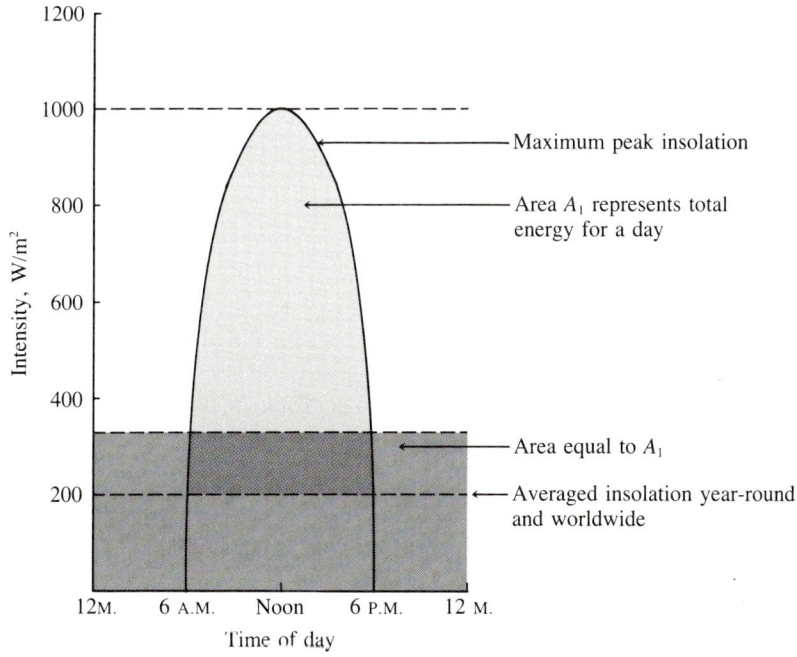

2. The area of 10 percent efficient solar cells under maximum terrestrial insolation required for a peak output of 1000 MW$_{pk}$, equivalent to the output of a typical nuclear or large coal-fired power plant, is 10^7 m^2, an area 1.9 mi on a side, or about 0.0001 percent of the total area of the United States.
3. The roof area of a typical single-family house in the United States is around 110 m^2, roughly 1200 ft^2.
4. The average annual electrical-energy consumption for a four-person household in a two-story single residence having a 1700-ft^2 living area is around 7150 kWh, corresponding to an average power utilization of about 815 W throughout the year (OTA, 1978, p. 699).
5. The efficiencies of commercial solar cells in 1980 ranged from about 10 to 15 percent. Typical costs of solar cells purchased in fairly large quantities

were around $10/W_{pk}$ in early 1980. Costs are predicted to fall to about $0.70 (1980 U.S. dollars) by 1985.
6. The cost of a single-crystal silicon solar cells made by conventional techniques is composed of three nearly equal parts: one-third is the cost of the silicon, one-third is the cost of fabricating the individual cells, and one-third is the cost of interconnecting cells and assembling arrays or modules.
7. The nationwide average cost of electricity in the United States in mid-1981 was 5.4¢/kWh. Prices for households, commercial users, and industrial users varied from values around 1¢/kWh in Seattle (1.2, 1.2, and 0.8¢, respectively) up to almost 14¢/kWh in New York (14.7, 14.0, and 11.5¢, respectively).
8. Balance-of-system costs are predicted to increase the cost of solar electric output by a factor of approximately 2.3 times that of the solar cell modules themselves.
9. Solar cells are expected to function for at least 20 years after installation.
10. Large-volume manufacture of electronic or mechanical parts typically results in prices dropping to about a third of the price for individually marketed units.

when solar cells will become economically practical. The key findings are that dispersed application sites will become practical first, and that feasibility will come in the mid-1980's *provided* the cost of solar cells continues to fall toward the levels taken as the DOE goals (Table 1.1), *and provided* the balance-of-system (BOS) costs can be reduced significantly in the next few years.

8.2 COST ANALYSES FOR PHOTOVOLTAIC POWER

It is useful to deal with the costs in three different ranges of power level:

- Residential user, 1- to 10-kW$_{pk}$ level
- Intermediate user, commercial and light industrial enterprises, 100 kW$_{pk}$ to several MW$_{pk}$
- Central power station, 100 MW$_{pk}$ to several thousand MW$_{pk}$

This division is useful because each of these users has quite different financial characteristics and constraints, as well as different electricity-demand schedules, knowledge and perception of the value of time to pay back an investment, and so on.

The dependence of the level and cost of PV power upon the characteristics of the application and the user are for the most part intuitively clear. The cost falls as the cost per watt of the solar cell modules drops, and the cost increases with increasing recurring costs of operating and maintaining the system. If the array efficiency decreases because of aging or is otherwise reduced by underutilization during some months when the load is light, this will increase the effective cost of the power produced and consumed. The greater the insolation, the lower will be the cost per watt produced, since a given array will have a greater output for the same acquisition price. Finally, the cost of assembling the funds to purchase the array will increase the apparent cost of the PV system itself.

These dependencies are expressed in Eq. (8.1.1) for the levelized cost of PV-produced power.

$$E = 10^5 \frac{(C)(F)}{(U)(S)} + OM \qquad (8.1.1)$$

In this equation E, the cost in cents per kWh, is *levelized,* meaning it is the ratio of the total costs incurred throughout the life of the system, divided by the number of peak kilowatthours of energy the system produces in its useful life (see Fig. 8.1). C is the installed cost of the system (array and associated equipment) in \$/W_{pk}. S is the ratio of the energy in kWh generated annually to the power rating of the system in kW_{pk}; under ideal conditions S equals the number of hours in a year, 8760, divided by about 5 to account for the ratio of daily average to peak insolation. The utilization factor U, accounting for factors that tend to reduce system output or its value, is discussed below. OM is the operation and maintenance cost, in cents per kilowatthour; it includes charges for cleaning panels, making repairs, and the like. This cost is assumed to be the same during each year the system operates. The factor 10^5 corrects for the mixed units used in the equation. Finally, F is the fixed-charge rate that represents the cost of financing the system. F equals the sum of the annual capital-related charges divided by the initial installed cost of the equipment.

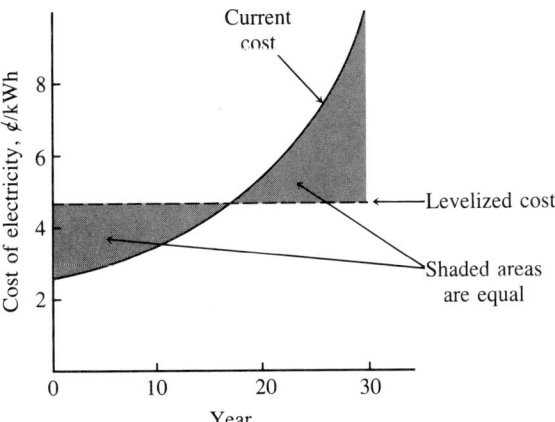

Figure 8.1 Levelized and actual cost of energy. The two shaded areas are equal.

Let us consider the utilization coefficient U and the fixed-charge rate in somewhat more detail. The utilization coefficient U accounts for factors that tend to reduce output of the system, including reduction of optical efficiency due to accumulation of dirt and to degradation of the surfaces through which the light must pass (possibly 8 percent averaged over the life of the system), power-conditioning losses (likely no more than 10 percent), and losses arising in storing and then retrieving energy (if 40 percent of the energy is stored in batteries and only 75 percent of that can be retrieved, then U is reduced by 10 percent). In addition, U is affected by changes in system efficiency due to operating at temperatures higher than the design value (efficiency and U are reduced), or lower than the design value (U increased). Finally, sale of energy back to the utility at too low a price can be thought of as a form of under-utilization, reducing U and increasing the effective cost of the output from the PV system. It has been estimated that these factors combined will lead to utilization coefficient values (U) around 0.7, and thus increase the cost of PV power by about 40 percent.

The fixed-charge rate F varies widely for the three different classes of user. For the residential consumer having a rooftop PV system that is treated as an integral part of the residence when the residence is purchased, F involves the amount of the down payment, interest, taxes, and insurance on the PV system, with correction for the tax bracket of the owner and hence possible reduction of interest through tax deductions. Of course, any tax credits allowed for installation of a PV system subtract directly from the cost term C. Typical fixed-charge rates F assumed in calculations of the economic feasibility of PV systems (OTA, 1978) have ranged from about 0.09 to 0.16.

For the intermediate-level application, there is a wide diversity of fixed-charge rates F, the values being generally low for schools and residential condominium or townhouse clusters and high for most commercial enterprises such as shopping centers. For these users there are large differences in the match between insolation and electricity demand. For utilities the fixed-charge rate is also high, including factors such as higher costs of borrowing, substantial business taxes, and payments to stockholders if the utility is publicly owned.

Balance-of-System Costs

Balance-of-system costs can be put in three categories:

1. *"Hard" BOS costs* are incurred only once for the purchase of equipment or for construction and installation. They include costs of power-conditioning and energy-storage equipment, land, foundations and support structures, lightning protection, transportation, and installation.
2. *"Soft" BOS costs* produce no visible equipment or structures and are

amenable to reduction through organizational, rather than technological, changes. These costs cover marketing and distribution, warranty of system, project management and architect and engineer fees, and interest during construction.
3. *Recurring BOS costs* result from operation expenses, maintenance, and replacement of elements that fail.

The DOE system price goal for 1986, when solar cells are assumed to cost no more than $0.70/W$_{pk}$, is $1.60/W$_{pk}$, requiring the total BOS costs to be only $0.90/W$_{pk}$. Present BOS costs are more than ten times this amount, and so a number of cost-reducing measures must be explored.

One can hope for technological improvements to reduce hard BOS costs, an example being the development of improved batteries or redox energy-storage systems that are less expensive than the lead-acid batteries now used. In certain applications, some of the hard BOS costs can be avoided altogether. For example, if the dc output of arrays can be used directly, the cost of inverters is avoided, saving at least $1/W$_{pk}$ at present prices. In some applications, as noted earlier, energy storage may not be necessary, saving at present prices at least $0.12/W$_{pk}$. Costs for land, foundations, and support structures may be relatively insignificant in those applications where PV cells can be mounted on existing structures, such as the rooftops of existing buildings. Further, installation costs can be minimized if standardized modules are designed for installation by relatively unskilled laborers.

Standardization of modules would also help reduce part of the soft BOS costs in custom installations where project management and architect and engineer fees may be substantial. If there is a boom in PV use, marketing and distribution costs might be minimized if the systems were distributed through existing large merchandising outlets, such as those that sell television, radio, and audio systems. Public willingness to invest in PV systems may also increase significantly if a warranty on the entire system is offered either by distributors or by vertically integrated companies that provide all components of the PV systems rather than just the cells themselves. Several such companies already exist and are listed in Appendix 8.

Cost Comparisons

The results of the DOE analysis (Clorfeine, 1980) for residential, intermediate, and central power station installations in three United States cities appear in Table 8.1. Anticipated costs of PV-produced electricity are compared with those expected for conventionally produced electricity (assumed to increase at a rate 3 percent higher than the inflation rate). Note that if the cell and system price goals are met, both in sunny Phoenix, where electricity from conventional sources is relatively inexpensive, and in far less sunny

Table 8.1 Tentative PV system price goals (1980 dollars)

User (date)	Location	Predicted conventional energy price, ¢/kW	System price goal, $/kWh	PV power price if goal is met ¢/kW
Residential (1986)	Phoenix	5.7	1.60	5.2
	Miami	5.5	1.60	6.9
	Boston	9.4	1.60	8.7
Selected intermediate (1986)	Phoenix	6.4	1.60	5.5
	Miami	7.0	1.60	7.3
	Boston	8.0	1.60	9.2
Central power station (1990)	Phoenix	All differ depending on whether baseload, intermediate, or peaking supply	1.10–1.30	4.2–4.8
	Miami		1.10–1.30	5.5–6.4
	Boston		1.10–1.30	7.0–8.1

Boston, where costs of transport drive up the prices of conventional power, the levelized prices of PV electricity in 1986 would be about 90 percent of those of the conventional sources of electricity for the residential consumer. Table 8.1 also shows that, at least for the Phoenix area, PV electricity will become less expensive than conventional power for the intermediate level consumer under the many conditions assumed. The central-power-station costs tabulated must be compared with actual utility costs to determine how much later they may become competitive.

Several other studies have produced similar conclusions but with interesting differences of detail. In analyzing an energy-efficient house located in Phoenix, Hammond (1979) has shown the importance played by the source of back-up power. This house had an assumed annual energy consumption of 7110 kWh/yr and fully 45 percent of that was consumed by evaporative coolers (rather than refrigeration systems) for air conditioning. The shares that these appliances required are of interest in themselves (Table 8.2). Clearly the economic feasibility of PV power depends on what conveniences the consumer regards as essential, since cooling for comfort and drying of clothes in an electric dryer consume more than half the electrical power in this house.

The simulated outputs through the year for two different-sized modules are plotted with the assumed demand in Fig. 8.2. The larger array meets nearly the entire peak load and it produces at other times far more power than is needed. The smaller module produces somewhat more power than is required during June, July, and August. The deficit is assumed to be supplied

Table 8.2 Power ratings and typical estimated annual energy consumption of electric household appliances

Appliance	Typical power rating, W	Typical annual energy consumption, kWh
Electric range with oven	12,000	700
Clothes dryer	4,500	1,000
Dishwasher	1,500	400
Microwave oven	1,450	190
Toaster	1,200	40
Air conditioner	1,100	1,400
Clothes iron	1,000	100
Coffee maker	1,000	110
Automatic clothes washer	700	100
Color television	500	500
Black-and-white television	350	350
Refrigerator/freezer with automatic defrost (16 ft^3)	–	1,800 1,230*
Refrigerator/freezer with manual defrost (12.5 ft^3)	–	1,500
Vacuum cleaner	300	50
Radio	100	90
Incandescent light bulbs	60–150 each	1,500
Fluorescent light bulbs	15–40 each	400
Electric clock	2	18

* Average value for 1979–80 energy-efficient refrigerator/freezer

by a backup source that would provide in those three months 18, 34, and 24 percent of the load, respectively. If the utility could have supplied the backup power in 1979 when the analysis was made, at then-current prices the cost could have been for the 30-yr life $0.26/kWh. If purchased in 1989 when solar cell modules are assumed to have dropped to no more than $0.70/$W_{pk}$ in 1980 dollars, the cost would be only about 40 percent of that of conventional sources. This analysis assumes that BOS costs drop as indicated by the DOE system goals, and that the utility can act as the source of backup power. If the back up source must be a diesel-driven generator at the load site, then the PV system will not be less expensive than conventional sources.

The interested reader will find in the OTA (1978) publication a detailed discussion of the issue of economic feasibility, and even a Fortran computer program developed for making cost projections. The entire energy needs (electricity, water heating, and space heating and cooling) are analyzed for the following types of energy consumers with assumed locations of Albuquerque, Boston, Ft. Worth, and Omaha:

- Single-family residence
- A 10-story, 196-unit high-rise apartment building

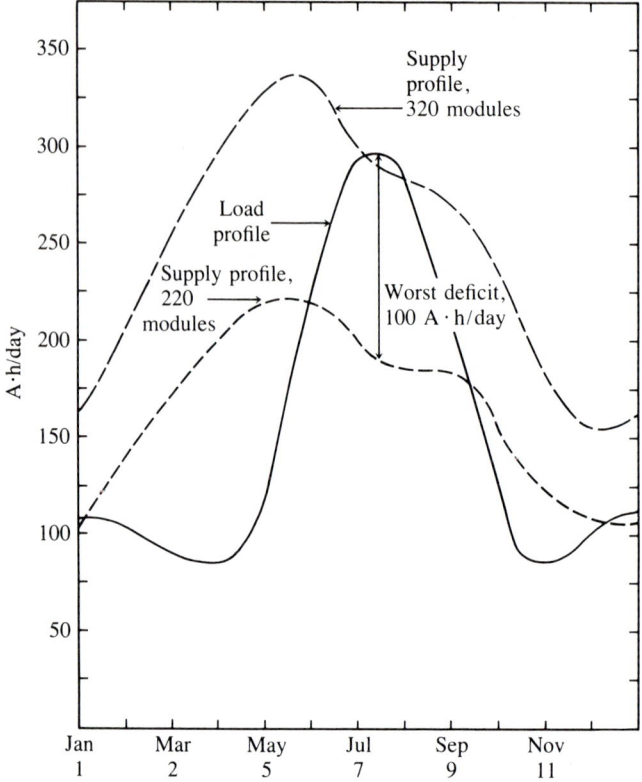

Figure 8.2 Simulated output for two differently sized PV arrays in Phoenix, Arizona, and assumed demand for a single residence. (*After Hammond, 1979.*)

- A 300,000-ft^2 shopping mall
- An entire residential community
- Various industrial installations needing electricity and process heat

The OTA study results generally show that PV systems can be competitive with conventional sources in residential and intermediate uses if the DOE price goals are met. Figure 8.3, on pp. 166–167, is one of several hundred case studies run by OTA, included here to show the level of detail necessary to yield predicted costs. The chief conclusions of the study for this house in Boston fitted with flat PV panels appear in part C of Fig. 8.3. They are that, if no financial incentives are offered for the installation of solar equipment, the levelized cost of PV electricity will be less than that for conventional sources even if a conservative estimate is made about the rate of escalation of electricity prices (the conservative assumption is that those prices rise at the rate of general inflation, which was assumed in this study to be only 5 percent). With steeper cost increases or with incentives such as a 20 percent solar investment tax credit, the savings to the user of the PV system are even greater.

Financial Incentives and Buy-back Policy

A variety of incentives may be available to encourage individuals to install PV systems. These include state and federal tax credits, low-interest loans, exclusion of PV equipment from the tax base on which property tax is figured, exemption from sales tax, and so on. As examples, in 1980 U.S. federal tax laws permitted an energy tax credit of 15 percent of costs up to a maximum credit of $300 for certain energy conservation items and 40 percent of costs, up to a maximum credit of $4000, for purchase of solar, geothermal, and wind renewable energy-source items. The State of California in 1980 offered a tax credit for individuals of up to $3000 per function (e.g., a PV system providing both electricity and hot water serves two functions, so a $6000 maximum credit applied), but only for taxpayers having an income below $30,000 for couples or $15,000 for single individuals. An electric utility company offered loans for the purchase of solar equipment at an interest rate about one-third the prevailing commercial rate.

The effects of these incentives can be appreciated by considering the changes they make in the life-cycle costs of a PV system (Carmichael et al., 1980). To simplify the example, we assume no buy-back of excess power by the local utility (this issue is considered next) and do not consider the cost of backup power bought from the utility. The yearly cost of the PV system is then

Yearly cost = (mortgage payment) + (increase in property tax)
+ (increase in insurance premium)
+ (operating and maintenance cost)
− (income tax savings for deducting interest
and taxes paid on the system)

The life-cycle costs of the system are the sum of the initial costs—down payment and sales tax—and the yearly costs, corrected for inflation and discounted to reflect the time value of money, for each year of system life. Table 8.3 (p. 168) shows the conditions assumed for a model residential PV system, and Table 8.4 shows the effects of various financial incentives on the life-cycle costs.

The data in Table 8.4 show that the major incentives are produced by exclusion of the PV system from the property tax base, credit on income tax, and the larger interest subsidies. The reader can get a better feeling for the contributions to the cost by verifying some of these figures; of particular

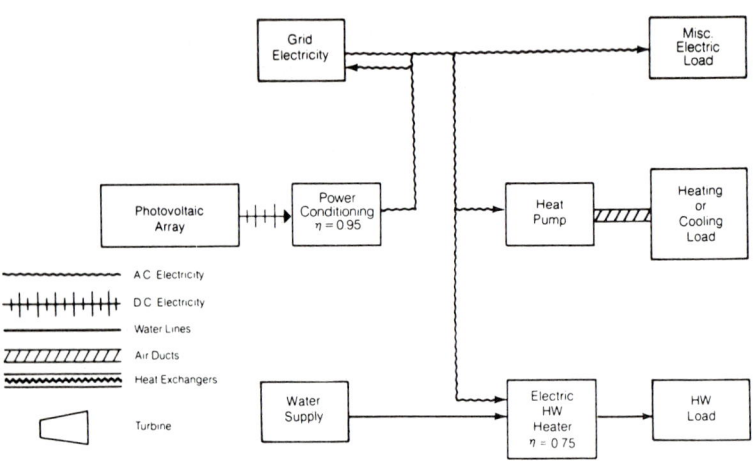

A. ITEMIZED COST OF COMPONENTS

Component	Size	Unit cost	First cost (incl. O&P)	Annual O&M	Life (yrs)
1. Electric heat pump	1.47 tons	800 $/ton	$1,180	$50	10
2. Ductwork	—	—	425	0	30
3. Electric hot water	40 gal.	$225 ea.	225	0	15
4. Air-cooled silicon PV (500 $/kW) ($\eta$ = 0.12)	59 m²	88 $/m²	*2,600	0	30
—Silicon array @ 60 $/m²			*2,600	0	15
—Shipping @ 2 $/m²					
—Installation @ 8 $/m²					
—25% overhead and profit					
5. Power conditioning	7 kW	114	800	8	30
6. Lightning protection	—	—	300	0	30
7. Extra insulation, storm doors and windows	—	—	981	0	30
TOTAL			$9,111	$58	

* ½ installed collector cost assumed replaced in 15 yrs., with total replacement in 30 yrs.

ANNUAL ENERGY FLOWS
(Conventional reference system is IF-2)

	Energy consumed by ref. system	Backup consumed w/ solar/conservation	Energy saved (% of total)
Net Electricity (bought-sold) (MWh/unit)	27.1	17.1	36.8
Fuel consumed onsite (MMBtu/unit)	0.	0.	0.
Total energy requirement (bbl crude equiv.)ᵃ	66.	42.	36.8
Electricity sold to grid annually (MWh,entire building)			3.8
Annual peak electricity demand (kW, entire building)			14.5

Figure 8.3 Sample output from analysis of a hypothetical residential PV system in Boston assuming $0.50/$W_{pk}$ silicon cells (*OTA, 1978*). In part C of this sample note that even without financial incentives the cost of conventional electricity is higher than that of PV-produced electricity in 1986 and beyond.

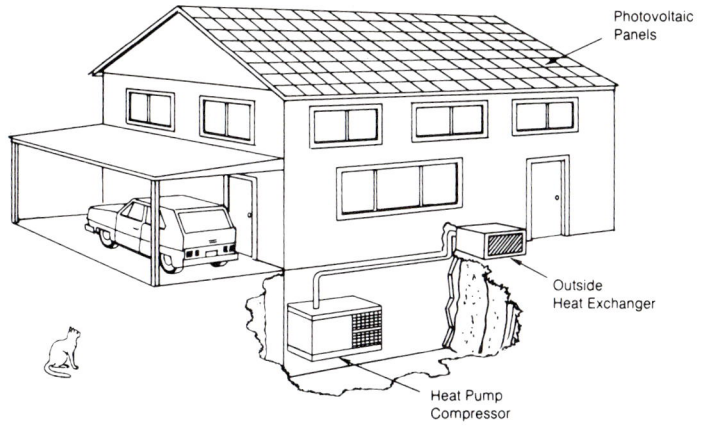

B. LEVELIZED MONTHLY COSTS PER UNIT TO CONSUMER (Dollars)[b,c]
(Conventional reference system is IF-2)

	Escalation of conventional energy costs		
	Constant real energy prices	Energy price escalation I	Energy price escalation II
1. 1976 Startup			
a. Costs using solar (conservation) system:			
Total with no incentives	248. (292.)	285. (328.)	433. (476.)
Total with 20% ITC	238. (283.)	274. (319.)	423. (468.)
Total with full incentives	225. (254.)	261. (291.)	409. (439.)
b. Costs using conventional reference system	226.	274.	467.
2. 1985 Startup[d]			
a. Costs using solar (conservation) system:			
(capital related costs)	99. (142.)	99. (142.)	99. (142.)
(operation & maintenance costs)	9. (9.)	9. (9.)	9. (9.)
(fuel bill)	0. (0.)	0. (0.)	0. (0.)
(electric bill)	141. (141.)	197. (197.)	428. (428.)
Total with no incentives	248. (292.)	305. (348.)	536. (579.)
Total with 20% ITC	238. (283.)	295. (340.)	525. (570.)
Total with full incentives	225. (254.)	281. (311.)	512. (541.)
b. Costs using conventional reference system	226.	300.	602.

C. EFFECTIVE COST OF ENERGY TO CONSUMER
(Conventional reference system is IF-2)

Levelized cost of solar energy or 'conservation' energy[c]	Type of incentives given		
	No incentives	20% ITC	Full incentives
$/MMBtu primary fuel	6.67 (11.10)	5.64 (10.22)	4.27 (7.28)
¢/kWh electricity	7.85 (13.07)	6.63 (12.03)	5.03 (8.57)

Levelized price paid for conventional energy[b,e]	Escalation of conventional energy costs		
	Constant real energy prices	Energy price escalation I	Energy price escalation II
$/MMBtu primary fuel	6.91	8.70	15.99
¢/kWh electricity	8.14	10.24	18.82

Figure 8.3 *(continued)*

Table 8.3 Characteristics of assumed residential PV system owned by home owner

System life	20 yr
Installed cost of system	$10,000
Inflation rate	8%
Interest rate	10%
Discount rate	10%
Operation and maintenance expense (% of installed cost)	1.5%
Insurance	0.3%
Property tax rate	2%
Sales tax	3%
Personal income tax bracket	25%
Down payment	10%
Tax benefit assumed, year after purchase	1 yr

Source: Carmichael et al., 1981.

Table 8.4 Effect of financial incentives on life-cycle cost of residential PV system
System has characteristics listed in Table 8.4 (Carmichael et al., 1981)

Case considered	Life-cycle cost, $	Reduction in life-cycle cost, $
Base case (parameters from Table 8.4)	13,758	—
Exemption from sales tax		
Sales tax rate 3%	13,458	300
Sales tax rate 5%	13,258	500
Interest subsidy		
Interest rate $9\frac{1}{2}$	13,539	219
Interest rate 5%	11,729	2,029
Interest rate 0%	10,125	3,633
Reduction in down payment		
5% down payment	13,672	86
No down payment	13,587	171
Exclusion from property tax base		
25% federal income tax bracket	11,454	2,304
Credit on federal income tax		
40% credit	10,112	3,636
55% credit	8,758	5,000

interest is the importance of the running expenses other than interest—property tax, operation and maintenance expense, and insurance. Incidentally, at this writing it appears that the tax credits for solar installations in the United States are likely to be abolished for fiscal reasons.

Turning to power buy-back policy, a portion (Section 210) of the U.S. Public Utilities Regulatory Policies Act of 1978 (PURPA) establishes important rules governing the flow of power between an electric utility and small power producers (Federal Register, 1980).

Prior to its enactment, a cogenerator or small producer of power wishing to sell excess power to the utility and buy power when it was needed might face three obstacles: (1) The utility was not obliged to purchase the power at an appropriate rate, (2) the utility might charge discriminatorily high rates for backup power supplied to the small producer, and (3) the small producer might be considered an electric utility and thus be subjected to quite burdensome regulations.

PURPA solves the third problem by defining suitable exemptions. The most significant provisions are these:

- ". . . electric utilities must purchase electric energy and capacity made available by qualifying cogenerators and small power producers at a rate reflecting the cost that the purchasing utility can avoid as a result . . . rather then generating an equivalent amount of energy itself."
- The electric utilities must furnish data concerning present and future costs of energy and capacity on their systems, so the small power producer can estimate the avoided costs.
- The utilities must furnish electric energy to qualifying facilities on a nondiscriminatory basis, and at a rate that is just and reasonable.

The first rule means that the *incremental* cost for a utility to generate an additional kilowatt-hour of energy at a specific time of day will be the price paid to the small producer who supplies the utility with a kilowatt-hour at that time. For power supplied at times of peak system loading, then, the price paid to the small producer may exceed the average price of energy charged by the utility, because at such times utilities use diesel generators to meet peak demands, at a cost higher than that incurred in operating more efficient baseload generators (see Fig. 8.4). Thus the conventional wisdom about buy-back policy—that power will be bought back at a "wholesale" rate significantly below the "retail" rate the utility charges its customers—is wrong. Table 8.5 shows the schedule of purchase prices for one utility during a period in 1980 when the average residential ("retail") rate was $0.04770/kWh. As the data in the table show, as much as a 20 percent premium will be paid for power supplied during peak system loading.

The effect of buy-back policy on the cost of PV power can be illustrated by an example. Suppose that in order to meet most of the annual needs a PV

170 ECONOMICS OF PHOTOVOLTAIC POWER

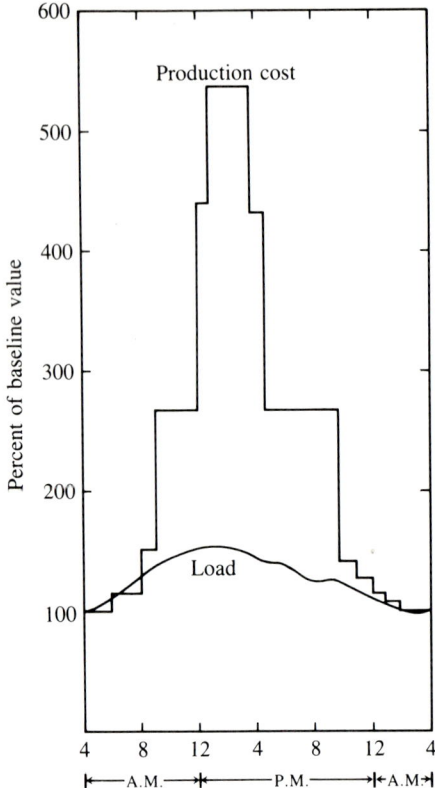

Figure 8.4 Summer daily demand pattern of typical utility. As the load increases above the minimum baseload value to its maximum, about 60 percent higher, the utility obtains power from more expensive generators and production cost increases by more than 400 percent. (*After EPRI Journal, Oct. 1981.*)

Table 8.5 Prices for buy-back of power from cogenerators and small power producers
May 1 through July 31, 1980 (Pacific Gas and Electric Company)

1. Standard weighted average price: 4.994 ¢/kWh
2. Optional time-of-delivery prices (producer pays for additional time-of-day metering equipment required):

   On-peak (12:30–6.30 P.M. weekdays)   5.675

   Partial-peak (8:30 A.M.–12:30 P.M. and
   6:30 P.M.–10:30 P.M. weekdays, and
   8:30 A.M.–10:30 P.M. Saturdays)   5.459

   Off-peak (10:30 P.M.–8:30 A.M., weekdays
   and Saturdays; all of Sunday)   4.700

system is slightly over-sized, so that through the year it produces 10 percent more power than can be used at the site. The utilization factor would be 0.90 in this case. Let us now consider two different utility buy-back policies, assuming for simplicity that the retail price of utility power just equals that of power from a PV system. If the utility were to buy power at only 50 percent of the retail price it charges, the utilization factor would be 0.95. If instead the excess power were produced at times of peak demand when the utility was paying 1.2 times its average retail price for electricity fed back to the grid, the utilization factor would become 1.02 (10 percent of power fed back at a 20 percent premium price). Thus the result of buy-back at this peak rate would be to lower the price of the PV power by from 7 percent (compared with 50 percent buy-back price) to 13 percent (compared with wasting the 10 percent excess power).

8.3 UTILITY ISSUES IN INDUSTRIALIZED COUNTRIES

Devoted proponents of solar energy may want to cut their ties to the utilities and "go it alone" when PV systems are available at costs somewhat lower than today's. Most people will likely choose to retain their utility connection, however, buying less energy than before and enjoying the benefits of system stability, backup power during periods of little insolation, and the sale of power to the utility when the solar units are producing more than is needed. It is important, therefore, to consider the issues that on-site PV systems will raise for the utilities.

To condense the vast amount of material on this topic we shall describe a possible scenario as it might be played in 1986 and beyond. The reader interested in more detail should start with the OTA (1978) volumes and the many references they contain on technical, economic, and legal issues involving utilities and on-site equipment.

Hypothetical Scenario for a PV-Powered Future

Let us assume that because of developments around the world, the DOE price goals for both solar cells and for BOS costs have been met. What might be the typical relationship between the utility and the resident of a building fitted with a PV system? What is the likely attitude of the utility regarding the increasing number of such on-site systems? What problems are likely to arise? Answers have been suggested by studies made by industry, government, and by the utilities themselves or their collective industry research organizations, such as the Electric Power Research Institute (EPRI) in the United States. Here is what we might see when the price goals have been met and on-site PV equipment is readily obtainable.

Since the reactions of those utilities that examined PV power by the early 1980s were generally positive, we find that the utilities welcome the growth of these on-site facilities. The standardized modular PV systems can be installed quickly, permitting the generating capacity to grow rapidly in response to increasing demand. The on-site systems are controlled by the utility, through signals sent over the grid to inexpensive electronic controllers at each site. This centralized control permits the utility to minimize the cost of backup energy supplied, since the utility alone can be aware of and responsive to the *marginal* cost of electricity. There are several different pricing schemes in use, having been gradually introduced around the world since the 1970s. Time-of-day pricing makes electricity furnished during times of peak demand more expensive than off-peak electricity, to reflect the higher cost of providing intermediate and peak-load power from diesel-driven generators. Control of appliances and equipment using large amounts of energy is lodged in the utility control room, from which the utility can transmit a signal over the electricity-distribution system to activate electronic switches at each customer's site. The resident may manually override this control, which can turn off electric clothes dryers and water-heating equipment, but the override option is seldom exercised because residents find that postponing the washing and heating can greatly reduce the utility bills. Water heated during off-peak hours is stored in well-insulated tanks for later use, as was being done already in the 1970s in residences in many European countries. Individuals who want still lower rates may opt for truly *interruptible* service, in which delivery of power from the utility can be terminated entirely for hours during periods of short supply.

In return for these additional system complexities, the utility has a demand curve that is considerably smoother than those typical of earlier times, so that large, efficient central baseload plants can generate a greater fraction of the utility power at low cost. In contrast with the strategy used by the operator of a typical stand-alone PV system, the utility does not store the electricity generated during most sunny hours, but instead distributes it to other users on the system. The excess PV-generated power supplied to the utility on weekends when the demand is low is stored in large load-leveling and peak-shaving storage batteries owned by the utility and located near the load centers to reduce distribution losses.

Consistent with the finding in the 1970s with windpowered generators was the observation that on-site PV generators feeding power back onto the grid do not cause problems because of transients or harmonics. No system stability problems arise even when 20 or 30 percent of the power on the grid comes from on-site PV generators. It had been feared that during system outages the supposedly dead grid wires might actually have thousands of volts across them because of on-site power flowing backward through what are normally step-*down* transformers. This hazard was eliminated simply by including an automatic cutoff switch at each site to open the connection to the utility when the central source ceased to function.

Credit for selling power back in some localities where the buy-back rate equals the rate at which electricity is sold requires no additional equipment since the ordinary integrating watt-hour meter runs either forward or backward and so registers the difference between the power received and that supplied. In other localities, a differential pricing scheme tied to time of day is in effect. An inexpensive electronic device registers the usage at the correct rate, totals the amount, and then, upon being interrogated electronically, transmits to the utility the information needed for billing.

Most consumers prefer to have the utility own and maintain the on-site equipment since they feel that raising the capital to buy the equipment outright is more burdensome than paying periodically to use utility-owned equipment—and to their surprise they find that study shows that the cost of the power produced is not materially different.

Whether events develop generally as outlined in this scenario depends upon technological development, and on the individual decisions and perceptions of literally millions of people. There appear to be no institutional, legal, or resource barriers that will prevent its becoming reality. The ultimate outcome depends on the achievable costs of the solar cell modules and the balance-of-system components.

8.4 ISSUES IN DEVELOPING COUNTRIES

On-site PV systems should be ideal for the developing nations where electric utility grids hardly exist. The systems are small and far less expensive to purchase than large central power stations, and they begin producing power within days or weeks after the start of their construction. The modularity of PV systems permits gradual, easily managed expansion as demand grows. The installations do not require imported fuel that must be transported long distances and whose price is subject to unpredictable increases. Thus the country is not required to commit itself to a form of energy production it may not be able to sustain because of increases in fuel prices, and this fact should encourage banks to loan money for PV systems.

The potential market in developing countries is huge, as there are more than 500 million people living in villages without any electric power. Life in these villages would be improved by the installation of even a small local generating system—an improvement that might reduce the drive for migration to crowded large cities in the often disappointing search for better conditions. A backup power source might not be needed in applications such as water pumping; if a backup source is required, it might be a diesel generator used only a few days each year and so requiring very little fuel.

Manufacture of some PV arrays could be done within many of the developing nations, possibly with unskilled labor and simply equipment. Because the cost of electricity in many parts of the world is several times higher than that in the industrialized nations, economic feasibility of PV systems will occur first in the developing nations as PV costs fall. Manufacturers might thus find that the market for PV technology—finished systems and manufacturing facilities—matures earliest in the developing countries and helps to support the growth of facilities for supplying photovoltaics to the industrialized nations.

8.5 SUMMARY

Solar cell systems will first compete economically at locations where the cost of conventionally produced electricity is highest—at remote locations, and at the load end of power-transmission systems. The levelized cost of PV power is sensitive to the consumer's financial and power-usage characteristics. Tax incentives and favorable utility buyback policies can greatly reduce the cost of PV-produced power. Studies show that PV power will be cheaper than conventionally produced power in the United States in the mid-1980s, provided the DOE cost goals for cells *and* balance-of-system components are met.

REFERENCES

Carmichael, D., et al. (1981), Institutional Issues—Major Issues and Their Impact on Widespread Use of Photovoltaic Systems, Record, 15th IEEE Photovoltaic Spec. Conf., 280–286.
Clorfeine, A. S. (1980), Economic Feasibility of Photovoltaic Energy Systems, Record, 14th IEEE Photovoltaic Spec. Conf., 986–989.
EPRI (1981), *Electric Power Research Institute Journal*, October.
Federal Register (1980), Public Utility Regulatory Policies Act of 1978 (PURPA), Vol. 45, no. 38, 12214–12237, 25 February.
Hammond, B. (1979), Solar Photovoltaic Power for Residential Use, ASTM Gas Turbine and Solar Energy Conf., San Diego, 12-15 March.
OTA (1978), Application of Solar Technology to Today's Energy Needs, Office of Technology Assessment, U.S. Government Printing Office, 2 vols. June, September.

PROBLEMS

8.1 *PV-powered auto* How long could a 25-hp vehicle operate on the electrical energy produced in a 24-h period by a 15 percent efficient PV system? (Assume insolation averages 200 W/m², the collector area is 1 × 4 m², and the motor-battery total efficiency is 70 percent.)

8.2 *Levelized cost* What is the levelized cost of electricity over a 20-yr period if the cost inflates from its initial value of 4¢/kWh at an 8 percent annual rate?

8.3 *Financial incentives* Identify what factors in Eq. (8.1.1) each of the financial incentives in Table 8.4 affects.

8.4 *Factors affecting economic viability* Examine some of the factors affecting economic viability of PV-generated electricity in your area. Is annual insolation high, average, or low? Is cost of utility-supplied electricity high, average, or low? If it is high or low, what factors account for this? Is time-of-day pricing in effect in your area? Do any of your major electrical energy demands occur when insolation is highest? (Air conditioning for comfort is an example for which energy storage and its associated cost may be avoided.) Does your electric utility have an energy buy-back policy? At what rates? Are area-related costs unusually high or low where you are?

CHAPTER NINE
ADVANCED MATERIAL PREPARATION AND PROCESSING TECHNIQUES

Forming and processing thin self-supporting semiconductor sheets for low-cost, high volume manufacture of cells

CHAPTER TEN
THIN-FILM AND UNCONVENTIONAL CELL MATERIALS

Low-cost cells from amorphous and evaporated semiconductors, and thin films formed on temporary substrates

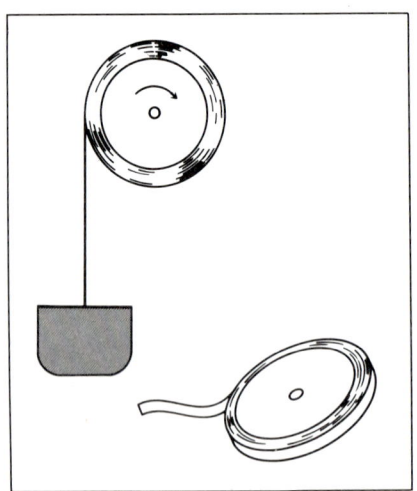
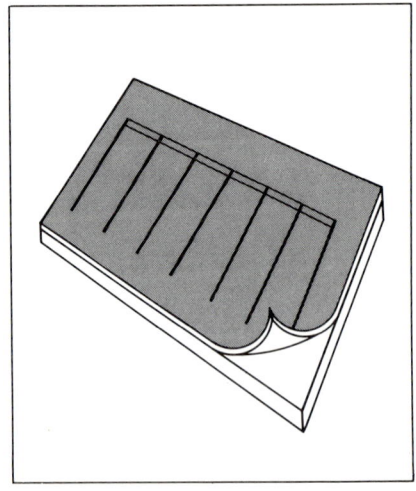

PART THREE

SOLAR CELL IMPROVEMENTS

CHAPTER ELEVEN
VARIATIONS IN CELL STRUCTURE

Multiple junction and other novel cell structures mostly for high efficiency

CHAPTER TWELVE
UNCONVENTIONAL CELL SYSTEMS

Cells employing photochemistry for energy storage, novel spectral manipulations for high efficiency, and the solar power satellite

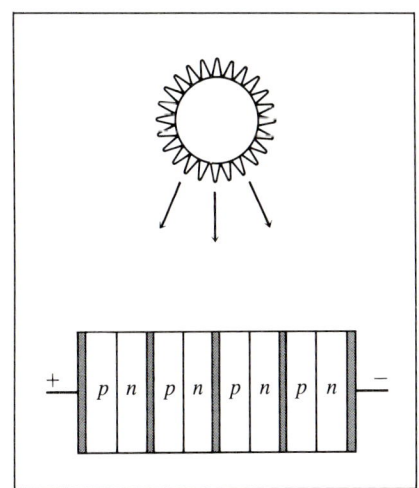

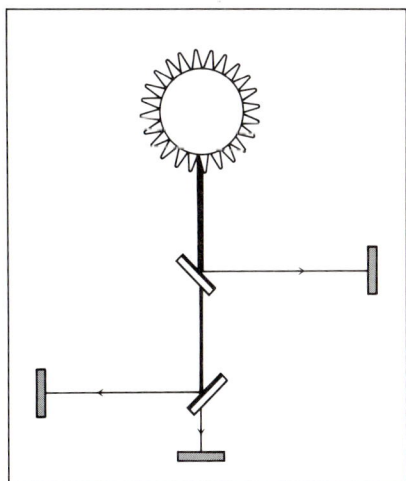

CHAPTER

NINE
ADVANCED CELL PROCESSING TECHNIQUES

CHAPTER OUTLINE

9.1 SOLIDIFICATION AND THERMAL ACTIVATION
9.2 FORMING THIN SELF-SUPPORTING SEMICONDUCTOR RIBBONS AND SHEETS
9.3 FORMING A SEMICONDUCTOR LAYER ON A SUBSTRATE
9.4 USE OF ION IMPLANTATION, LASERS, AND ELECTRON BEAMS
9.5 OPTICAL TRANSMISSION, CONTACTS, AND ENCAPSULANTS
9.6 CONTINUOUS PROCESSING IN AN AUTOMATED FACTORY
9.7 SUMMARY
REFERENCES
PROBLEMS

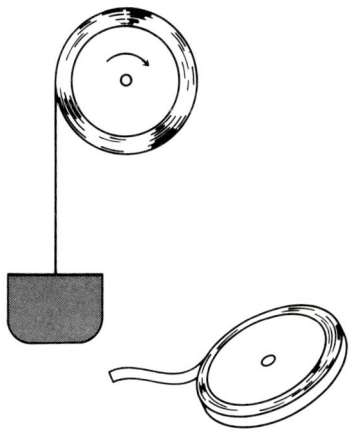

180 ADVANCED CELL PROCESSING TECHNIQUES

The motivations for investigating new cell processing techniques are reducing cell costs, increasing rates of cell production, and reducing the energy used in cell manufacture.

Conventional silicon cells have been made by separate and distinct manufacturing steps, as described in Chap. 4: growing the single-crystal semiconductor, cutting it into wafers, doping the wafers, forming contacts, and assembling entire arrays. With the newer fabrication methods, the demarcations between processes are blurred, as suggested by Fig. 9.1. The principal question to ask about the newer processes is, "What cell efficiency can be achieved?" The loss of carriers at imperfections in the cell material made by these methods is generally greater than in conventional cells, so the efficiency is usually lower. One must decide whether the advantages of lower cost and higher production rate compensate adequately for the lower efficiency.

In this chapter we describe promising new processes for making solar cells. These methods are grouped as follows: growing thin self-supporting semiconductor ribbons and sheets that do not require wafering; forming a semiconductor layer on a substrate; using ion implantation, lasers, and electron beams for doping, annealing, and stimulating grain growth; and achieving high optical transmission, putting on contacts, and encapsulating cells. Before discussing these topics we consider some general principles about solidification and defects in crystals, as these topics are particularly relevant in the newer methods of making cells. We conclude the chapter with a description of a hypothetical automated solar cell factory.

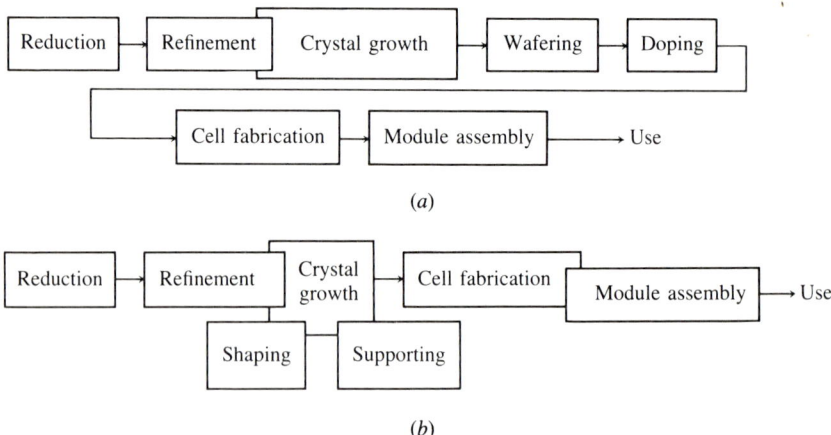

Figure 9.1 Conventional (*a*) and advanced cell processing (*b*). Some of the individual steps of conventional processing occur in a single step in the advanced processes, of which silicon ribbon growth and forming cadmium sulfide cells on float glass are examples.

9.1 SOLIDIFICATION AND THERMAL ACTIVATION

Solidification

The most common way of forming a solid is to freeze a molten mass of the material. When a molten elemental metal or semiconductor cools slowly, its temperature changes with time as shown by curve *A* in Fig. 9.2. The temperature falls until it reaches a definite value called the *freezing temperature* for that material, at which it remains for some time before beginning to fall again. At the freezing temperature, if heat is being removed from the volume rapidly, freezing will occur simultaneously at many different places in the liquid volume. Nuclei of solidifying material that reach a critical size will continue to grow as more atoms attach themselves. The latent heat given off in the freezing process flows to the melt, permitting it to remain at a constant temperature in spite of the removal of heat from its boundaries. The individual nuclei eventually become large enough to contact each other, forming a polycrystalline solid with grain boundaries.

Very different conditions are used when one makes large single crystals: A seed crystal is placed in contact with the melt for orientation, cooling is slow, and a thermal gradient is established between the melt and the solid at whose surface atoms are attaching themselves.

In a two-component melt in which both constituents are soluble in the liquid but insoluble in the solid that forms, one can show from the Gibbs phase rule that the cooling curve will have the form of curve *B* of Fig. 9.2. One of the atomic components begins to freeze out first, robbing the liquid phase of that component and changing the composition of the melt as the temperature continues to fall.

Thermal Activation

In the solidification processes that result in nearly perfect single crystals, atoms arriving at the surface of the growing solid must be able to reach the proper locations to continue the regular crystal structure. In a number of

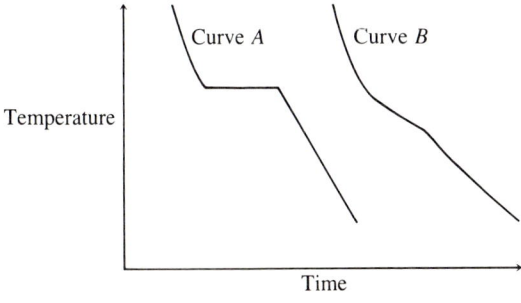

Figure 9.2 Cooling curves for an elemental solid (*A*) and for a compound solid (*B*).

182 ADVANCED CELL PROCESSING TECHNIQUES

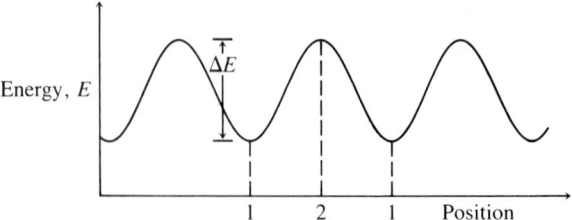

Figure 9.3 Energy of atom as a function of position in a regular crystalline solid, assumed one-dimensional for clarity.

processes that appear important for the next generation of low-cost solar cells, thermal energy is supplied to permit atoms to move to the crystallographically correct locations.

As examples, consider the growth of a layer of crystalline semiconductor upon another crystalline semiconductor. If the underlying semiconductor (A) and the semiconductor that is to form the layer (B) have the same crystal structure and nearly the same interatomic spacings, growth may be achieved from a melt (solid A in contact with molten semiconductor B), from a vapor (solid A in contact with vapor containing constituent B in some form), or by evaporation (constituent B evaporated onto solid A). These processes of epitaxial growth are known respectively as *liquid-phase* epitaxy (LPE), *vapor-phase* epitaxy (VPE), and, when used to form compounds, the evaporative process is called *molecular-beam* epitaxy (MBE). VPE is a special case of *chemical vapor deposition* (CVD), which is employed to deposit an amorphous or polycrystalline layer on a substrate.

In order for the atoms arriving at the surface to lodge permanently at the proper sites—those associated with minimum energy—the atoms must be able to move to those sites over the surface of the solid. One can represent the energy of an atom on the surface of a one-dimensional solid by a sequence of regularly spaced hills and valleys (Fig. 9.3). Heating the solid can provide atoms on the surface with enough energy to surmount the barriers and move over the surface to the lower energy sites. It is even possible to heat a disordered solid layer on a dissimilar or similar crystalline solid substrate so as to produce an ordered crystalline layer (*solid-phase* epitaxy, SPE).

Single-Crystal Growth by Czochralski, Float-Zone, and Bridgman Techniques

To form single-crystal silicon ingots, either the Czochralski (CZ) or the float-zone (FZ) solidification process is employed. In the CZ process, a small oriented piece of single-crystal silicon used as a seed is put into contact with the surface of the molten silicon in a crucible that is rotated slowly. Silicon from the melt freezes on the seed, which is slowly raised, resulting in the formation of a cylindrical single-crystal ingot from 7.5 to 12.5 cm in diameter and up to a meter long. Some purification also occurs during CZ growth, as can be seen from the phase diagram of Fig. 9.4. If constituent B is silicon and A is an impurity, the first solid to freeze will contain much less of impurity A than does the melt in contact with the solid that is forming.

In the float-zone method, a molten zone is passed through a relatively pure silicon ingot, causing a redistribution of impurities much as in the case of CZ growth. Because impurity concentrations are generally higher in the liquid, passage of thin molten zones through an ingot having a 0.01 percent impurity content causes impurities to be swept to one end of the ingot, leaving in the middle a region having an impurity content as small as one part in 10^{10}. Finally, the cylindrical ingot is cut with saws into round wafers about 250 μm thick.

Another important single-crystal growth technique is the Bridgman process, in which an oriented seed crystal is located at one end of a crucible containing the liquid, relatively pure semiconductor. Solidification occurs as the crucible or boat passes slowly through a region where the temperature is lowered to a value below the freezing temperature of the solid. This method (or CZ growth) is typically used for the growth and partial purification of GaAs. In growing GaAs by the Bridgman process one may start with liquid Ga in a sealed chamber that also contains a source of As. At the growth temperature, which is near the melting point of GaAs, 1240°C, the As source produces a vapor of As atoms at one atmosphere pressure, and stoichiometry of the GaAs results.

Growth must be relatively slow if single crystals are to result. The growth

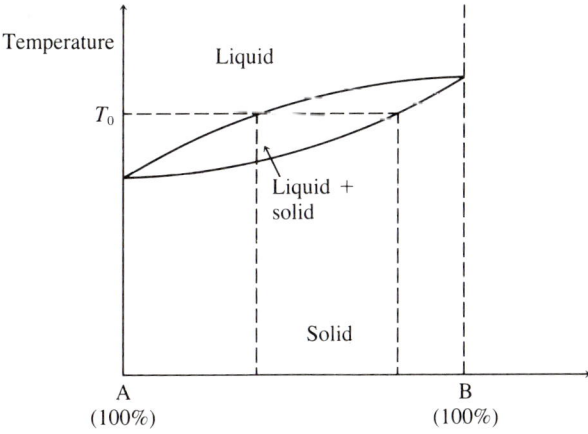

Figure 9.4 Hypothetical phase diagram for a two-component system. The constituents are labeled A and B.

rates for the CZ, FZ, and Bridgman techniques are typically on the order of centimeters per hour.

Polycrystalline Cells

The trade-offs of simpler and faster fabrication methods are well illustrated by the important example of polycrystalline solar cells. Polycrystalline silicon cells are made from square wafers of polycrystalline silicon cut from a solid that cooled from a melt. The grains in these cells have diameters of up to several millimeters, but the cells are otherwise similar to single crystal *pn*-junction Si cells. They have a diffused or implanted front layer, a metal front contact grid and a solid back electrode. Their square shape enables one to fill an array completely with these cells rather than leaving inactive areas as when one uses circular cells. Efficiencies up to 17 percent have been achieved (Storti, 1981). To correct for the loss of some carriers due to recombination at the grain boundaries, it has been possible to diffuse impurities along grain boundaries and thence a small distance into the individual grains to make the boundary regions serve to collect photo-generated current. This approach is shown in Fig. 9.5.

Improved Silicon Purification

The impact of improved processing in a particularly energy-intensive step in cell fabrication can be seen from two examples of ways of obtaining purer metallurgical grade silicon.

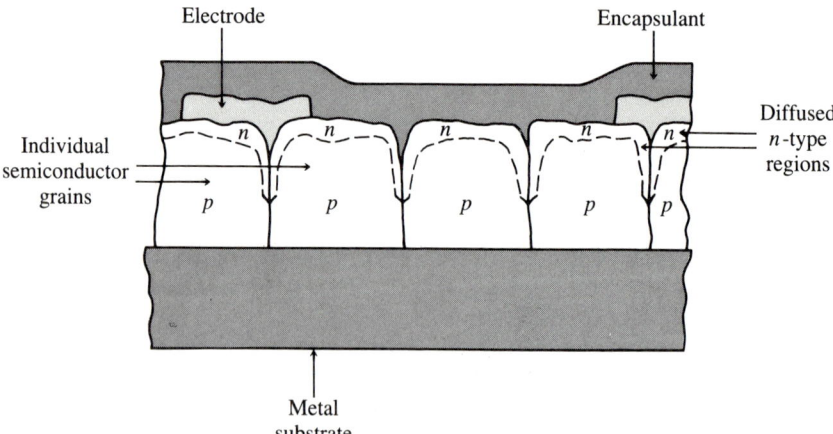

Figure 9.5 Cross-section of polycrystalline cell having diffused *n*-type regions partially surrounding each grain to collect carriers.

In Sec. 4.2.1 we showed that much of the energy payback time of conventional Si cell manufacture accrues in the silicon-refinement stage. It is possible to reduce the energy required there by using purer starting materials—naturally occurring quartzite pebbles having 99.5 percent SiO_2 content and refined charcoal that has been exposed to a halogen gas at high temperature for several hours (Lindmayer et al., 1977). When the molten silicon is cooled, unidirectional solidification can be employed and so some purification occurs then because of the segregation of impurities in the melt. If a polycrystalline starting material is acceptable for cell manufacture, a Czochralski process with a high pull rate can be used, resulting in further energy savings. The estimated cumulative payback time for this purification/refinement process is only 0.43 yr instead of the 2.84 yr cited earlier for the conventional reduction, refinement, and crystal-growth processes (these times do not include the energy expended in other steps in the cell fabrication process).

The energy payback time for these three processes can be lowered further to an estimated 0.18 years with a silicon fluoride polymer transport purification process (Lindmayer et al., 1977). Metallurgical-grade Si is exposed (at 1100°C) to SiF_4, producing gaseous SiF_2, which is then polymerized by chilling (to −45°C). Subsequent heating (to 400°C) produces amorphous Si and gaseous silicon fluorides, which escape and are recycled. Additional heating above 730°C causes the Si to crystallize and drives off all remaining silicon fluorides. Purification occurs at each stage of this process, making it quite economical in terms of the energy expended.

9.2 FORMING THIN SELF-SUPPORTING SEMICONDUCTOR RIBBONS AND SHEETS

Ingenious techniques exist for producing semiconductor plates thin enough to be made directly into solar cells without wafering. The desired thickness for silicon is from 0.1 to 0.3 mm. Single-crystal material is preferred for highest efficiency, although polycrystalline material is usable, particularly if the grains are large and their boundaries are mostly oriented parallel to the current-flow direction, perpendicular to the plane of the sheet. Furthermore, means are being developed for converting polycrystalline ribbons to single-crystal form, and for enlarging grains substantially in polycrystalline ribbons or sheets. One may also form a thin semiconducting sheet on a temporary substrate such as graphite or molybdenum and then remove the sheet for use in a cell. A dramatic new example of the use of the temporary substrate is the *rapid quenching process*—quenching at rates of up to a million degrees per second—originally used to form amorphous metal alloys (Chaudhari et al., 1980). Here molten silicon is sprayed onto a rapidly rotating, cooled metal cylinder, where it crystallizes immediately to form a ribbon that moves tangentially away from the cylinder at high speed.

Low cost and large-scale utilization of solar cells demand high growth rates from any candidate processes, along with reasonable cell efficiencies. These are somewhat contradictory requirements since fast growth often results in imperfect crystals having numerous defects at which recombination can occur. Another problem is controlling contamination. If molten or very hot solid silicon is shaped by dies or rollers, reaction with those parts can produce both serious contamination of the semiconductor and deterioration of the apparatus. With silicon formed by condensation of a vapor at a solid surface, avoiding contamination may require the use of diffusion-barrier layers of metals having small diffusion coefficients for impurities present in the substrates.

Characteristics of some of the most promising of the ribbon and sheet processes are illustrated in Fig. 9.6 and summarized in Table 9.1. These are the *edge-defined film-fed* (EFG) ribbon and the *dendritic web* growth processes, the high-speed *rapid quenching* process, and several other processes where silicon is formed on a temporary substrate. The *ribbon-to-ribbon* (RTR) regrowth process employing laser heating enlarges the grains in polycrystalline material (see Sec. 9.4). Other interesting variations that have been suggested include forming doped silicon filaments from which one can weave a "cloth" having photovoltaic response, forming silicon layers on a glass substrate using the intermediary of an aluminum-silicon eutectic to avoid melting of the glass, and making tiny single-crystal doped silicon microspheres for use in a photoelectrochemical solar cell (described in Chap. 12).

The *edge-defined film-fed growth* process for silicon-ribbon growth (Ravi et al., 1975) was developed by Mobil-Tyco from its original use in growing sapphire tubes for high-pressure lamps. It involves pulling one or more Si ribbons 150 to 300 μm thick and up to 5 to 10 cm wide through a die mounted on a chamber containing molten Si (Fig. 9.6*a*). Growth in a desired direction is begun by the use of a seed first put into contact with the silicon melt. Interestingly, such thin silicon is flexible enough to be rolled like movie film onto a reel. Pull rates of 5 cm/min have been achieved and solar cell panels

---

Figure 9.6 Methods for producing self-supporting silicon ribbons and sheets: (*a*) Edge-defined film-fed growth (EFG) process. Multiple ribbons can be pulled from an apparatus having many dies and a single reservoir of molten silicon. (*b*) Stepanov ribbon process. (*c*) Dendritic web growth process for making Si ribbons up to 5 cm wide. [(*a*), (*b*), *and* (*c*) *from U.S. Patent 4,121,965.*] (*d*) Pulling broad Si sheet horizontally from the surface of a melt (schematic). (*e*) Schematic illustration of high-velocity rapid-quenching process, in which Si ribbon emerges at speeds up to tens of meters per second.

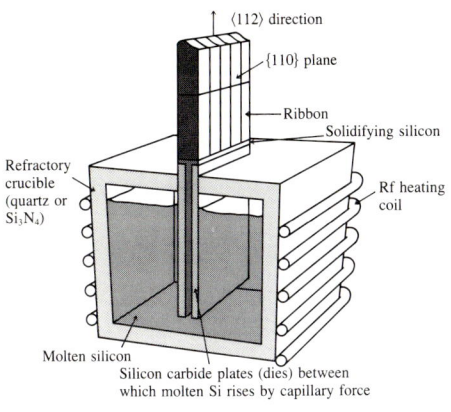

(a)

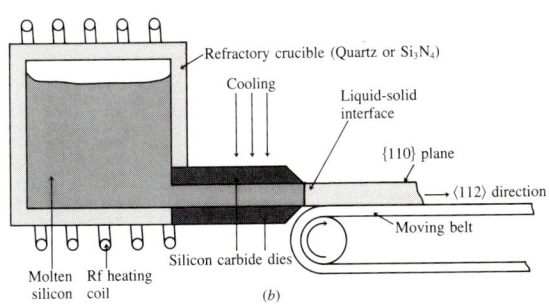

(b)

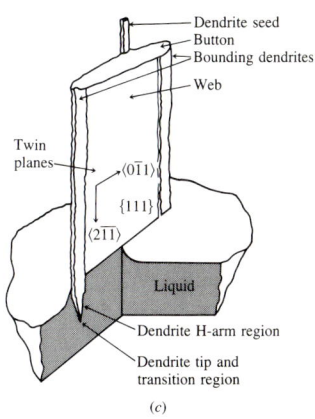

(c)

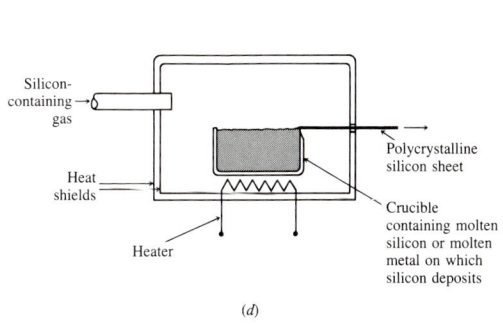

(d)

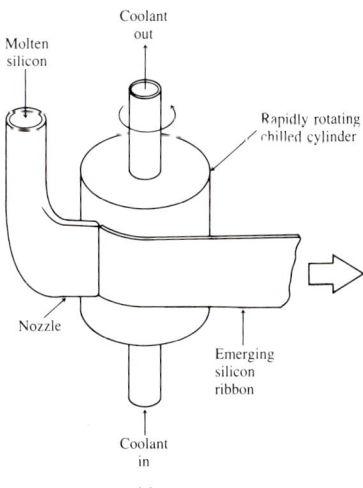

(e)

187

Table 9.1 Characteristics of some rapid silicon-growth techniques
Where two values are given, upper ones are best reported values and lower are typical results

Process	Experimental cell efficiency, %	Growth rate, cm/min	Width, cm	Thickness, μm	Continuous growth results
EFG ribbon	13–14 10–11	5 2.5–4	5–10	150–200 250–300	five 30-m lengths, 15 hrs
Dendritic web	15.5 12–13	7 3–4	4.7 3.5–4	50 150	3- to 4-m lengths, 17 hrs
Horizontal ribbon growth	9–10	41.5	5 1–3	200–350 400–2000	Greater than 5-m length
Low-angle silicon sheet	?	20–60	0.8–2.5	300–1000	0.75-m length
Rapid-quenching ribbon	5–8	1.8×10^5 (30 m/s)	0.1–1 (to 5 with double roller)	20–200	
Ribbon-to-ribbon regrowth	13 11–12	9 2.5–5	7.5 3.5	100 150–200	30-cm lengths
Silicon-on-ceramic	10.1 9	15 4–5	10 5	100 200	50-cm lengths

Source: Surek, 1980; Arai, 1980.

having 11 percent overall efficiency have been made from EFG ribbon. The economics of using such ribbon in an automated continuous solar cell factory are summarized in Sec. 9.6. Chief technical difficulties are the dissolving of the SiC die in the highly reactive molten Si, and the formation of crystal twins in the ribbon, which increases recombination at the boundaries. In the closely related *Stepanov process* (Fig. 9.6b), silicon is pulled horizontally from a hole formed by a die in the bottom of a heated boat, again resulting in essentially single-crystal ribbon having final thickness (Leipold, 1978).

Process parameters and characteristics of cells made from EFG material are listed in Table 9.1, along with those of the other ribbon, sheet, and layer processes described in this chapter. No clearly "best" process is yet evident since none of the processes receives consistently high marks in all the important characteristics required for high-quantity production of cells producing low-cost electricity.

In the *dendritic web* process (Fig. 9.6c), no die is used. A thin Si web is grown between two dendrites formed by pulling on a suitable seed initially in contact with a pool of molten Si. Growth rates up to 7 cm/min are obtained. Close control of temperatures and pulling rate is required, and the web surface is quite smooth. The advantages of the dendritic web process can be obtained with less difficult thermal control by growing a web between two graphite or quartz filaments pulled through a silicon melt (*edge-supported* pulling process) (Ciszek and Hurd, 1980). Both the dendritic web and the EFG processes have been put into at least pilot production, and in 1980 a formal arrangement was announced between a web manufacturer, Westinghouse, and two large electric power companies to begin production and evaluation on a commercial scale.

Horizontal ribbon growth by pulling a sheet horizontally from the surface of a melt permits somewhat higher rates of production than provided by the preceding ribbon techniques (Fig. 9.6d). This is because the large surface-to-volume ratio of the silicon permits more rapid removal of heat by radiation and convection from the emerging sheet. This process should have fewer problems associated with dissolving of the die than does the EFG process.

A very high rate of ribbon production is achieved by the *rapid quenching* process (Fig. 9.6e). Molten Si at 1500°C is ejected in air from a quartz tube onto a chilled copper cylinder 40 cm in diameter spinning at 100 r/min (Arai et al., 1980). The Si quenches as a polycrystalline ribbon with 20- to 30-μm diameter columnar grains whose axes are perpendicular to the ribbon plane. Doping to form a surface layer of different conductivity type can be accomplished, it is claimed, by ejecting an impurity-containing vapor onto the emerging ribbon downstream from where it leaves the cylinder. Ribbon widths from 1 to 50 mm, thicknesses from 20 to 200 μm, and linear velocities from 10 to 40 m/s are said to be possible. With a 20-m/s ribbon velocity—45 mi/h—one quenching apparatus could form 3 km^2 of 50-mm-wide ribbon in a year, corresponding to an impressive 300-MW$_{pk}$ output in AM1 sunlight if the cell efficiency were 10 percent. In early experiments 5 percent efficient cells were obtained, not by the vapor-doping method but by use of an n-type Si layer produced by chemical vapor deposition (CVD) on the p-type ribbons.

Another method for producing self-supporting polycrystalline ribbons is removing molten Si from the bottom of a heated container by squeezing it to the desired thickness between chilled rollers. Other workers have examined the feasibility of forming thin sheets by solidification on a temporary substrate such as a molybdenum sheet, in the case of both Si and GaAs. In one such process, the difference in thermal expansion coefficients of the 100- to 250-μm thick Si film and the Mo substrate causes the Si film to peel off upon cooling; after grain growth induced by the RTR process (Sec. 9.4), cells were made having AM1 efficiencies as high as 11.8 percent (Sarma and Rice, 1980). A process involving the cleavage of lateral epitaxial films for transfer to another substrate (the CLEFT process) is described in the next section, where formation of films on substrates is considered.

9.3 FORMING A SEMICONDUCTOR LAYER ON A SUBSTRATE

Single-crystal semiconductor layers grown slowly, under near thermal-equilibrium conditions, on a regular oriented solid substrate, may have excellent electrical and optical properties. Such layers may be used as the body of the solar cell, as a front-contact layer of opposite conductivity type, or as a layer for surface passivation and optical transmission as in the case of the GaAlAs layer on a crystalline GaAs cell.

Traditional processes for forming single-crystal layers by epitaxy—growth upon an oriented substrate—by the LPE or VPE processes have involved naturally occurring single-crystal substrates. Growth rates are typically between a micron per minute and a micron per hour, and growth temperatures range from hundreds to a thousand or more degrees centigrade. The change in properties at the interface is abrupt, and usually a fairly close match of lattice dimensions is required of the substrate and the layer, along with similarity of the thermal-expansion coefficients of those materials.

The molecular beam epitaxy (MBE) process is like VPE but with important differences. An ultra-high vacuum in the 10^{-10} Torr range is used for MBE, and evaporated atoms from heated sources form the growing layers, resulting in a fairly low growth rate on the order of tens of angstroms per minute. With MBE, semiconductors having compositions that cannot be achieved in thermal-equilibrium growth have been made, opening the possibility of tailoring electrical and optical properties, including the energy gap. Although applicable to solar cell work, it is doubtful that MBE will be inexpensive enough to be used extensively for making nonconcentrator solar cells; the process may be useful for making small, high-efficiency multicolor concentrator cells, however.

Epitaxy on artificially regular surfaces formed by texturing photolithographically with a periodic pattern written with an electron beam has recently been achieved and christened "graphoepitaxy" (Geis et al., 1979). Single-crystal silicon has been formed by repeatedly laser-heating a 300-nm-thick amorphous silicon film grown by CVD on an amorphous silica substrate into which 100-nm-deep grooves having a 3.8-μm period were made by etching. It may be possible to mass-produce inexpensive textured surfaces for graphoepitaxy by embossing or injection-molding from a master, as is done with the videodisc—phonograph-recordlike disks having micron-sized depressions that store video information for later readout by a scanned low-power laser beam.

Very promising early results have been obtained with the so-called CLEFT process (McClelland et al., 1980), in which a thin semiconductor film is grown on and then cleaved from a substrate. CLEFT is an acronym for *cleavage of lateral epitaxial films for transfer*. The method can be used with silicon, gallium arsenide, and probably other semiconductors. In one experiment, a carbonized photoresist mask having narrow, widely spaced stripe openings was formed on a (110) GaAs substrate that was placed in a VPE chamber. Epitaxial growth starts in the openings and then proceeds laterally over the mask, producing a continuous single-crystal GaAs film that can be cleaved from the reusable substrate. A 17 percent efficient GaAs single-junction cell has been made by the CLEFT process.

The many ways of depositing *polycrystalline semiconductor layers* that may be used for solar cells include the following:

• Evaporation from a boat of the semiconductor, possibly heated by an electron beam to maintain purity, onto an inexpensive inert substrate. An example is the evaporation of GaAs from a boat containing the compound onto graphite-coated molybdenum sheets.

• CVD-formed semiconductor layers on an inexpensive substrate. GaAs films 10 μm thick formed on tungsten-coated graphite have been used to make 8.5 percent efficient MOS cells (Chu et al., 1981).

• Columnar polycrystalline deposits having 10-μm-diameter grains of both Si and GaAs have been made by evaporation at rates up to 1 μm/min onto aluminum films, which form the back ohmic contact, on a glazed alumina ceramic support. An electron beam (E-gun) is used to heat the Si source ingot; in the GaAs deposition, the Ga source is heated with an electron beam while As is evaporated from a resistance-heated quartz crucible. Small grain size and contamination from the substrate remain problems.

- Evaporation onto thin flexible plastic substrates, such as mylar and kapton, of semiconductors such as CdS was demonstrated and used years ago in space cells that were unrolled in orbit, suggesting the possibility of making and installing terrestrial cells in a similar way.
- Polycrystalline deposits of elemental and compound semiconductors such as Ge, Si, and ZnO, as well as AR coatings, can be made by sputtering: in a chamber containing an ionized inert gas at low pressure, bombardment by the ions knocks atoms from a source (a plate or powder of semiconductor) so that they ultimately deposit on the desired substrate located in the chamber. Ionization may be produced by a dc or rf power supply. Deposition rates may exceed 20 μm/hr, and very large continuous-flow magnetron sputtering systems exist (for putting partially reflective coatings on sheets of window glass).
- "Silicon-on-ceramic" cells made from films of Si produced by dipping a substrate such as mullite ($3Al_2O_3 \cdot 3SiO_2$) into molten Si have been evaluated and found to yield up to 10.5 percent efficiency, as quoted by Feucht (1980).
- In the *ribbon-against-drop* process, a 300-μm-thick graphite ribbon coated with 5-μm-thick carbon deposited by pyrolysis is drawn at 7 cm/min through a close-fitting die in the bottom of a chamber of molten Si. The result is a coating of the ribbon by a polycrystalline Si layer less than 100 μm thick. The meniscus at the die prevents leakage from the chamber. Cells having 8 percent AM1 efficiency have been made this way.
- Manufacture of cadmium sulfide cells on glass emerging at the end of the production line of a float-glass plant has been reported by several groups; the processes include evaporation and spraying.
- Semiconductor films potentially useful for solar cells have been made by electrochemical deposition (electroplating) either from an aqueous or a nonaqueous electrolyte. These electrochemical processes are very efficient, can produce oriented granules, are fast with 1-μm/min and higher deposition rates for metals, and are suitable for large-scale, high-volume production. To date, reported efficiencies have been disappointingly low.
- A fascinating variant on these methods for producing uniform continuous coatings is the silicon shot process, described more fully in Chap. 12 in connection with photoelectrochemical cells. Tiny silicon spheres made like lead shot in a cooling tower are embedded in a plastic sheet and exposed to sunlight while in contact with a liquid electrolyte.

Amorphous semiconductor films have been produced in plasma discharge systems, for use in solar cells that are expected to be very inexpensive. Because its properties are quite different from those of either single-crystal or polycrystalline forms of the same semiconductor, amorphous material and means for its production will be discussed in Chap. 10, where unconventional solar cell materials are considered.

9.4 USE OF ION IMPLANTATION, LASERS, AND ELECTRON BEAMS

Doping semiconductors by ion implantation is a standard integrated-circuit-processing technique that is expected to be important in solar cell manufacture. Implanters operating with beam voltages from 10 to more than 300 keV are commercially available, permitting one to produce impurity densities up to about 10^{16} cm^{-2} in continous processing of Si wafers loaded in automatically handled cassettes. Throughputs of 250 to 300 4-in-diameter wafers per hour are possible, at costs around \$0.01 to 0.02/cm^2 (1980 dollars).

In addition to forming the front contact layer, ion implantation can be used to deposit a backside layer in back-surface-field (BSF) cells where minority carriers are deliberately reflected from the heavily doped back contact. Another potential use of implantation is in *gettering,* a process that can be used to increase minority-carrier lifetime in semiconductors by sweeping lifetime-killing impurities and certain crystal defects away from crucial regions of a semiconductor device. In gettering with impurities, a dopant is deposited to a high density by implantation (or by thermal diffusion) through the back surface of a cell. Typical dopants are P or Ar, at densitites around 5×10^{15} cm^{-2}. When the wafer is later heated in an inert atmosphere (for example, at 850°C in forming gas), interstitial impurity atoms and crystal defects migrate through the crystal and become trapped at the doped surface, where they may be removed with a chemical etchant or left where they do not affect device performance. An alternative gettering process involves simply damaging the back surface mechanically before heating. These processes are found to increase the lifetime markedly, and have been used commercially for recycling reject wafers from the IC industry.

Thermal energy is usually required for atoms in a solid to move about at an appreciable rate. In conventional semiconductor processing, heating occurs primarily by radiation from a resistive heater, held at constant temperature, surrounding a fused quartz furnace tube containing a high-purity gas. New methods of heating involving transient energy input to the semiconductor with intense light or high-density electron beams appear very attractive for solar cell manufacture.

Heating may be used for diffusing dopants, for healing the lattice damage caused by ion implantation, enabling grains of a polycrystalline solid to grow larger, relieving strains caused by unequal thermal expansion in a multilayer semiconductor structure, promoting chemical reactions such as oxidation and epitaxy, sintering surface layers, and even melting surface layers or particularly susceptible layers inside a solid. Using intense electron beams or light from a laser or an intense incoherent light source can be efficient—since the energy can be deposited just where it is needed—and fast, permitting a high throughput.

Two different regimes of heating with electron beams and light are distinguished: a *pulsed* regime involving brief pulses of high-enough energy density to cause melting of part of the irradiated solid, and a *scanned* regime characterized by a more gradual deposition of energy that may or may not be sufficient to cause melting but is adequate for producing rapid diffusion of atoms in the solid. In the pulsed mode, electron beam or Q-switched laser peak intensities up to 10 MW/cm^2 are used, so that surface temperatures rise in a few microseconds from the steady substrate temperature to the melting temperature. When melting occurs, in a layer about a micron thick, atoms mix rapidly in the melt, causing the concentration profile of implanted dopants to change from gaussian or near-gaussian shape to a uniform distribution, as shown in Fig. 9.7. At lower intensities in the scanning mode, a continuous electron beam or ion laser beam is swept at a rate of a few centimeters per second, electronically or mechanically respectively, over the surface of the semiconductor mounted on a stage heated to an average temperature around 300°C. Atomic rearrangement can take place (solid-state epitaxial regrowth), although melting does not occur, to activate impurities—

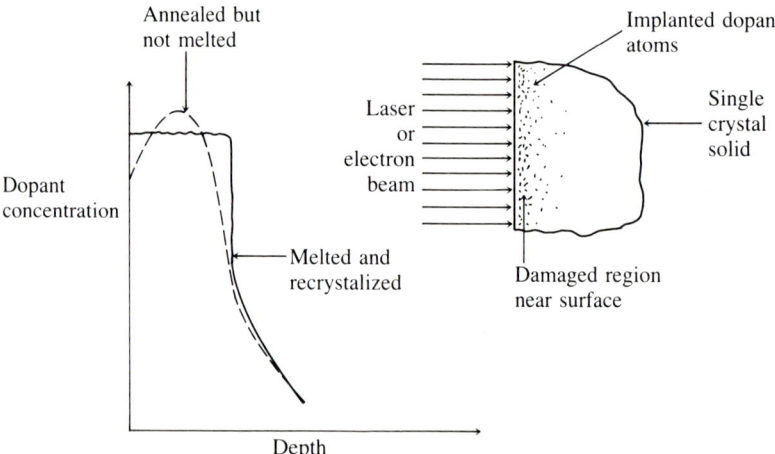

Figure 9.7 Dopant profiles obtained by ion implantation and laser- or electron-beam annealing. Inset shows distribution of atoms after implantation, with damaged surface layer and electrically inactive dopant atoms. Use of intense pulsed laser or electron-beam irradiation causes melting of the surface layer and uniform redistribution of dopants in the layer. Lower temperature heating also activates impurities and removes damage but does not redistribute the dopants, allowing the impurity profiles obtained by control of implant conditions to be retained.

allow them to move into substitutional sites where they ionize—and to heal lattice damage.

It has been shown that scanned electron beam and laser annealing of implants can activate implanted impurities as fully as does thermal annealing. The effects upon minority-carrier lifetime have not yet been clearly established. Electron beam systems convert primary power to beam power much more efficiently than do lasers, but require the semiconductor to be in a vacuum. Inclusion of a pulsed electron beam source or introduction of a laser beam through a window in an ion-implantation machine for annealing permits both processes to be done in a single pumpdown. With optical heating, the penetration depth can be controlled by the choice of wavelength, within practical limits based on available high-intensity light sources. The use of concentrated sunlight as a source is an intriguing but perhaps impractical idea, compatible with the "solar breeder" factory described in Sec. 9.6.

Recrystallization of ribbon by laser heating (the ribbon-to-ribbon, or RTR, process) is illustrated in Fig. 9.8. The growth of crystals having desired directions is promoted by heating, which permits atomic rearrangement but not actual melting, and a thinner, single-crystal ribbon emerges at the top of the recrystallization apparatus. Conversion of inexpensively deposited amor-

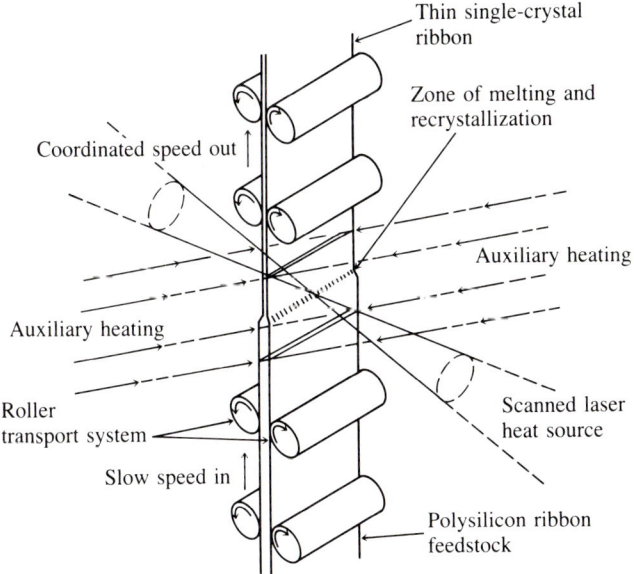

Figure 9.8 Ribbon-to-ribbon process for converting a polycrystalline silicon ribbon entering at bottom into thin, single-crystal ribbon by controlled laser heating. (*From Jet Propulsion Laboratory, California Institute of Technology, "Low Cost Solar Array Project Quarterly Report No. 9, April–June 1978," Pasadena, CA, 1979.*)

phous Si and GaAs into crystalline form has been demonstrated, as mentioned in Sec. 9.3, and as described in experiments with a graphite strip heater moved slowly laterally near the surface of an amorphous semiconductor (Tsaur et al., 1981). Successful grain enlargement and polycrystalline regrowth of Si and GaAs have been reported by many workers.

9.5 OPTICAL COATINGS, CONTACTS, AND ENCAPSULANTS

The topics of this section will be treated briefly since the aims of increasing production rate and lowering cost of optical coatings, contacts, and encapsulants appear to be attainable by fairly straightforward means. For example, presently most solar cell AR coatings are deposited on 20 or 30 wafers mounted together in an evacuated chamber by evaporation of oxides or fluorides from a heated filament or boat. For larger production rates, continous processing must be used, whether it be evaporation or magnetron sputtering, with cells moved automatically into and out of the coating chamber through air locks. Some alternative processes that might find use in optical coating are the use of sprayed or spun-on liquid coatings, in precisely controlled amounts, which are then fired to yield dense and durable coatings. Where glass sheets are used for encapsulation, chemical leaching of the surface is a fast way of removing material at the glass–air interface to make the change of index of refraction less abrupt and so increase optical transmission. Electrochemical means such as electroplating followed by anodic-oxidation (anodization) is a high-rate, low-cost process that has already been used experimentally to form AR coatings.

The dimensions of the solar cell electrodes are large, at least on non-concentrator cells, compared with the dimensions now common in ICs, permitting the use of processes akin to printing—electrode "ink" consisting of metal powders mixed with an organic binder and perhaps very small beads of low-melting-point glass (frit) is printed or silk-screened on the semiconductor and then fired. In Schottky-barrier cells, the careful control of the thickness of the broad area electrode to permit both optical transmission and current collection likely will require use of evaporation or magnetron sputtering combined with feedback from an electrical or optical thickness-monitoring system. A heavily doped semiconductor having a wide bandgap can serve as a transparent conducting electrode in Schottky-barrier cells. The outstanding example is indium-tin oxide (ITO), discussed in Sec. 10.3.

Encapsulation with either glass or compliant rubbery substances has proved most reliable in solar cell tests to date, although it is claimed that suitable varnishes have been found that protect the cadmium sulfide cells made on a float-glass substrate. For example, in marine solar arrays polyvinyl butyral has been used as a resilient encapsulant, and butyl rubber as an edge sealant. Glasses have been bonded electrostatically—by the field-assisted bonding process—to oxidized semiconductors and metals by applying briefly an electric field around 10 kV/cm to a sandwich consisting of glass in close contact with the semiconductor between electrodes (with the negative on the glass side) heated to several hundred degrees centigrade. It appears possible with some concentrator cell designs to encapsulate the cell almost entirely in glass, leaving open only the backside of the semiconductor for heat transfer. Solder glass frits which can be melted with intense infrared radiation may also be useful in encapsulating finished cells.

9.6 CONTINUOUS CELL PRODUCTION IN AN AUTOMATED FACTORY

Simulation with the SAMICS computer methodology for analyzing the manufacture of solar cells (see Appendix 7) has shown that an automated factory based on a silicon-ribbon technology can produce cells that meet the 1986 DOE cost goals—provided certain assumptions are valid.

A crucial assumption is that polycrystalline metallurgical-grade silicon (MG–Si) is available at $14/kg (1980 dollars). In large quantities, such as the 1000 metric tons that would be required annually for the 250 MW/yr factory discussed below, silicon made conventionally costs about $56/kg (1980 dollars). It is claimed that processes under development could result in prices ranging from $10/kg to $14/kg. The silicon cost is only one element in the overall array cost, so the estimated cost per watt for the processing described below to make 12 percent efficient arrays would only increase from $0.63/$W_{pk}$ with $10/kg silicon to $0.85/$W_{pk}$ if one had to use the $56/kg silicon available now (again, the 1986 DOE target is $0.70/$W_{pk}$).

The business assumptions are based on a conventional capitalistic framework with amortization of equipment facilities, a 100 percent overhead rate on direct costs, 20 percent return on invested capital, 9.25 percent interest rate on borrowed capital, and a 1:6 debt-to-equity ratio. Ordinary corporate taxation and normal tax credits are assumed. The factory, having a 250-MW annual capacity, would operate round-the-clock and all of its product is assumed to be sold.

Silicon ribbon producing 12 percent efficient encapsulated AM1 panels is assumed. Five 7.5-cm-wide EFG ribbons are pulled from each furnace at 7.5 cm/min. Other ribbon technologies such as the dendritic web or the high-speed rapid-quenching process would also be usable, of course, at possibly lower cost.

The processing assumed follows, and the estimated costs appear in Table 9.2. A p^+ back contact is made by applying and firing-in aluminum powder

Table 9.2 Annual cost in $/W_{pk}$ for process steps in hypothetical automated factory producing solar panels with continuous processes

Values in 1975 dollars. Encapsulated cells assumed 12% efficient; $10/kg silicon assumed

Process	% total value added	Value added	Capital costs	Direct labor	Materials and supplies	Utilities	Indirect expenses	Yield
Silicon preparation	~10	$0.043			$0.0434			
Sheet fabrication	~31	0.145	0.0630	0.0312	0.0140	0.0049	0.0320	0.800
Cell fabrication	~23							
p^+ Back		0.002	0.0010	0.0004	0.0002	0.0000	0.0005	0.998
Etch		0.010	0.0036	0.0018	0.0033	0.0000	0.0018	0.994
Ion implant		0.012	0.0067	0.0018	0.0000	0.0003	0.0032	0.998
Pulse anneal		0.006	0.0038	0.0004	0.0000	0.0003	0.0011	0.992
Back metallization		0.036	0.0108	0.0013	0.0206	0.0005	0.0030	0.980
Front metallization		0.036	0.0111	0.0013	0.0202	0.0005	0.0030	0.980
AR coating		0.009	0.0038	0.0018	0.0014	0.0002	0.0016	0.990
Interconnection, packaging, and testing	~36							
Interconnect		0.033	0.0104	0.0040	0.0135	0.0000	0.0054	0.999
Encapsulate and assemble		0.130	0.0368	0.0062	0.0750	0.0001	0.0120	0.999
Test		0.001	0.0003	0.0002	0.0000	0.0000	0.0002	0.980
Package		0.001	0.0002	0.0001	0.0002	0.0000	0.0001	0.9999
Total	100%	$0.464	0.1515	0.0505	$0.1918	0.0068	0.0639	—

Source: Aster, 1978.

on the back of the wafer. Before ion implantation and pulsed annealing of the front layer of the junction, a plasma etch removes the thin oxide that forms naturally on the wafer. Screen-printed silver paste forms the back and front contacts, and a sprayed antireflection coating is applied. Cells are interconnected into 1.2-m-long strings of 12 cells each, and then 16 such strings (192 cells) are connected into a 4-ft^2 module, and encapsulated between a 3-mm-thick float glass and a 0.005-in-thick mylar back layer, with ethylene vinyl acetate sheets between. There is an edge seal and an aluminum frame. Incidentally, the encapsulation materials, excepting cells and interconnects, would cost around $12/m^2 (1980 dollars). Modules are then tested and packaged. Moving belts would transport parts from one work station to the next for high throughput and low labor cost. Throughput at some stations would exceed 60 cells per minute, and the total processing time from semiconductor stock to module could be as short as two hours. Of the processes listed, cost estimates for the pulse annealing of the ion implant are not firm because this process, though shown to work, is not operated today on a commercial scale. Printing of electrodes and contacts also may require some further development for large-scale use.

Note that in Table 9.2 the estimated silicon cost is only about 10 percent of the final cost, and that sheet fabrication and final assembly of cells into encapsulated modules each account for about one third of the total added value. The sheet fabrication is the most expensive item in terms of both capital and labor requirements per watt of cell output; the encapsulation and assembly involve the highest material and supply costs. Energy payback times for the steps shown are estimated at about two months for the 12 percent cells, *excluding* the energy for constructing the factory and energy for silicon production. The amount of silicon used is only about a third of that required by an advanced ingot technology, causing a corresponding reduction of the estimated energy payback time down to slightly less than one year.

Solar Breeder Plant

One intriguing concept is a solar cell plant whose only direct energy input is sunlight incident on arrays of solar cells on its roof (Lindmayer et al., 1977)—the "solar breeder," a logical alternative to the nuclear version. Perhaps simply on principle, direct solar thermal input has been avoided in the design, although it could also be used. To handle the problem of intermittency of sunshine, two forms of storage are envisioned—electrical storage in batteries, and temporary storage of partially completed objects between manufacturing stations. Equipment such as large furnaces in the plant must be operated continuously, but other equipment could be run intermittently as energy supply and demand for particular parts fluctuate. Computer control of processing, now common in IC manufacture, will be particularly important in this case because of the need for responding appropriately to the fluctuating energy input.

9.7 SUMMARY

Process heat employed in making conventional solar cells contributes significantly to cell cost. Improved processing techniques aimed at achieving low costs involve reducing the heat required—through improved refinement and purification processing, and utilizing polycrystalline semiconductors. By using thin sheets of semiconductor formed directly rather than by sawing from an ingot, one can realize additional important savings of material, energy, and labor. Examples of the latter are the dendritic-web silicon-growth process, and the CLEFT process for growing single-crystal Si or GaAs sheets that are later separated mechanically from their growth substrates.

Recrystallization of rapidly grown polycrystalline layers should raise the efficiencies of the cells, and automated large-volume cell manufacturing and array assembly should help make solar cells competitive with conventional power sources.

REFERENCES

Arai, K. L., Tsuya, N., and Takeuchi, T. (1980), Ultra-High-Speed Growth of Silicon Ribbons for Solar Cells, Record, 14th IEEE Photovoltaic Spec. Conf., 31–35.

Chaudhari, P., Gressen, B. C., and Turnbull, D. (1980), Metallic Glasses, *Sci. Am.* Vol. 242, No. 4, 98ff, April.

Chu, S. S., Chu, T. L., Jiang, C. L., Loh, C. W., Stokes, E. D., and Yu, J. M. (1981), Thin-Film Gallium Arsenide Solar Cells with Reduced Film Thickness, Record, 15th IEEE Photovoltaic Spec. Conf., 1310–1315.

Ciszek, T. F., and Hurd, J. L. (1980), Proc. Symp. on Novel Silicon Growth Methods, Electrochem. Soc., May.

Feucht, D. L. (1981), Recent Progress in the Development of Advanced Solar Cells, Record, 15th IEEE Photovoltaic Spec. Conf., 648–653.

Geis, M. W., Flanders, D. C., and Smith, H. I. (1979), Crystallographic Orientation of Silicon on an Amorphous Substrate Using an Artificial Surface-Relief Grating and Laser Crystallization, *Appl. Phys. Lett.*, Vol. 35, No. 1, 71–74, 1 July.

Leipold, M. H. (1978), U.S. Pat. 4,121,965; 24 October.

Lindmayer, J., Wihl, M., and Scheinine, A. (1977), Energy Requirements for the Production of Silicon Solar Arrays, Report SX/111/3 (ERDA/JPL/954606-77/3), Solarex Corp., Rockville, Md., October.

McClelland, R. W., Bozler, C. O., and Fan, J. C. C. (1980), A Technique for Producing Epitaxial Films on Reusable Substrates, *Appl. Phys. Lett.*, Vol. 37, No. 6, 560–562, 15 September.

Ravi, K. V., Serreze, H. B., Bates, H. E., Morrison, A. D., Jewett, D. N., and Ho,

J. C. T. (1975), EFG Silicon Ribbon Solar Cells, Record, 11th IEEE Photovoltaic Spec. Conf., 280–289.

Sarma, K. R., and Rice, M. J., Jr. (1980), The Thermal Expansion Shear Separation (TESS) Technique for Producing Thin Self-Supporting Silicon Films for Low-Cost Solar Cells, *IEEE Trans. Elec. Dev.*, Vol. ED-27, No. 4, 651–656, April.

Storti, G. M. (1981), The Fabrication of a 17% AM1 Efficient Semicrystalline Silicon Solar Cell, Record, 15th IEEE Photovoltaic Spec. Conf., 442–443.

Tsaur, B. Y., Fan, J. C. C., Geis, M. W., Silverman, D. J., and Mountain, R. W. (1981), Improved Techniques for Growth of Large-Area Single-Crystal Si Sheets over SiO_2 Using Lateral Epitaxy by Seeded Solidification, *Appl. Phys. Lett.*, Vol. 39, No. 7, 561–563, 1 October.

PROBLEMS

9.1 *Thermal activation and impurity diffusion* The atomic diffusion constant D at any absolute temperature T for a given impurity in silicon can be written as $D = D_0 e^{-E_a/kT}$, where E_a is an activation energy, k is the Boltzmann constant, and D_0 is a constant that depends on the type of impurity. In a time t atoms will diffuse a mean distance $L = 2(Dt)^{1/2}$, where L is known as the atomic diffusion length. From the data below, determine how long it would take at 100°C for boron (used intentionally as a dopant) and copper (an unwanted contaminant) to diffuse entirely through a 100-μm-thick silicon solar cell, making it inoperable.

Impurity atom	Temperature, T	D, $\mu m^2/h$
Boron	1573 K (1300°C)	2.9
	1173 K (900°C)	3.2×10^{-4}
Copper	1573 K (1300°C)	6.4×10^7
	1173 K (900°C)	1.9×10^7

9.2 *Production rates, improved processes* Referring back to the example of Chap. 4 for meeting 10 percent of the U.S. electricity demand in the year 2000 with PV systems, calculate the number of EFG ribbon or dendritic web machines required to meet the annual production rate. Assume the best of the characteristics of these processes listed in Table 9.1. Compare with the value for the rapid-quenching process.

9.3 *Nonthermal processing methods* The nonthermal processes described in this chapter that use ion implantation, lasers, or electron beams tend to be more energy efficient than entirely thermal processes, such as CZ growth and thermal diffusion. Identify where input energy is lost in these different processes (i.e., fails to produce a useful result). What constraints on the wavelengths used in laser regrowth (Fig. 9.8) can you identify?

9.4 *Solar cell factory* Describe your visions for the most advanced silicon solar cell manufacturing plants that will be operational in 10 yr and in 20 yr. Be specific. You may also describe the raw materials and the finished products.

CHAPTER

TEN

THIN-FILM AND UNCONVENTIONAL CELL MATERIALS

CHAPTER OUTLINE

10.1 INTRODUCTION
10.2 AMORPHOUS SEMICONDUCTORS
10.3 MISCELLANEOUS MATERIALS FOR SOLAR CELL USE
10.4 PROSPECTS FOR MASS-PRODUCED THIN-FILM CELLS
10.5 SUMMARY
REFERENCES
PROBLEMS

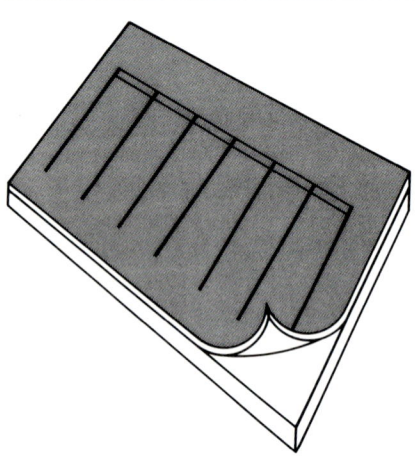

Of the many substances that exhibit the photovoltaic effect, cells having AM1 efficiencies greater than 15 percent have been made with only two materials—Si and GaAs. On the other hand, cells with AM1 efficiencies greater than 5 percent have been made utilizing more than a dozen different semiconductors, and one may well wonder whether there is one "best" semiconductor material for solar cells.

The problem is that the promise or potential of a given material for PV use is not determined solely by the *intrinsic* properties of the semiconductor, and so no simple figure of merit for PV use exists. Important factors affecting cell behavior and cost, though hard to quantify, include cell structure (*pn* junction or MIS, for example), form in which the material is used (such as single-crystal or polycrystalline), the effects of processing upon carrier lifetime, suitability for large-scale manufacture, stability under changing ambient conditions, and material availability and potential toxicity. As a result, many different materials have been and are being investigated.

10.1 INTRODUCTION

The most important physical characteristic of a candidate solar cell semiconductor is the energy gap, since it determines to what fraction of the solar spectrum the semiconductor can respond. The maximum theoretical con-

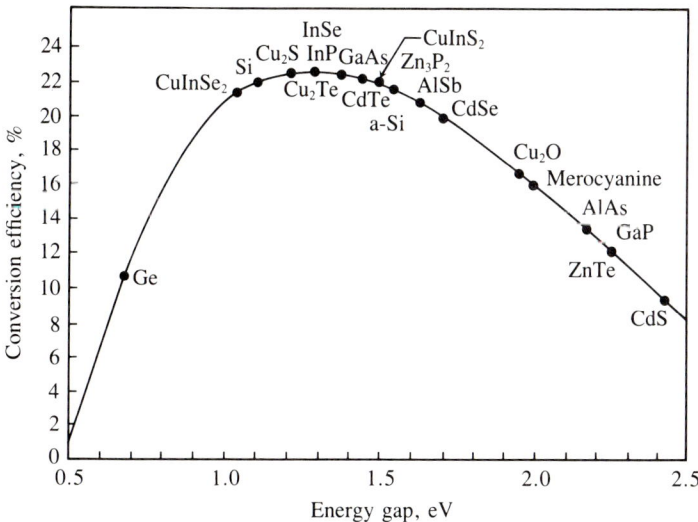

Figure 10.1 Maximum theoretical conversion efficiency vs. energy gap for solar cells in AM1 sunlight. Crystalline silicon is denoted Si, and amorphous silicon is a–Si. [*Barnett and Rothwarf (1980), as adapted from Prince (1955). © 1980 IEEE.*]

version efficiency is plotted in Fig. 10.1 versus semiconductor energy gap. A number of semiconductors of current interest for solar cell applications are identified in the figure. Of great importance also is the optical absorption coefficient and its dependence upon photon energy, plotted in Fig. 10.2 for many semiconductors being considered for use in thin-film nonconcentrator cells. The AM1.5 solar spectrum is also shown in Fig. 10.2; semiconductors having an energy gap less than 1.7 eV would permit utilization of a significant portion of that spectrum. See also Fig. 3.5.

The qualitative effects of many of the characteristics or properties of semiconductors are listed in Table 10.1, which applies mostly to pn-junction cells. A more restricted list of the properties of many semiconductors appears in Table 10.2 where only energy gap and selected electrical properties appear. Of the commercially available cells discussed in Chap. 4, Si is the most used, even though its energy gap is not optimal, because the material is relatively inexpensive and the technology of processing it is well developed.

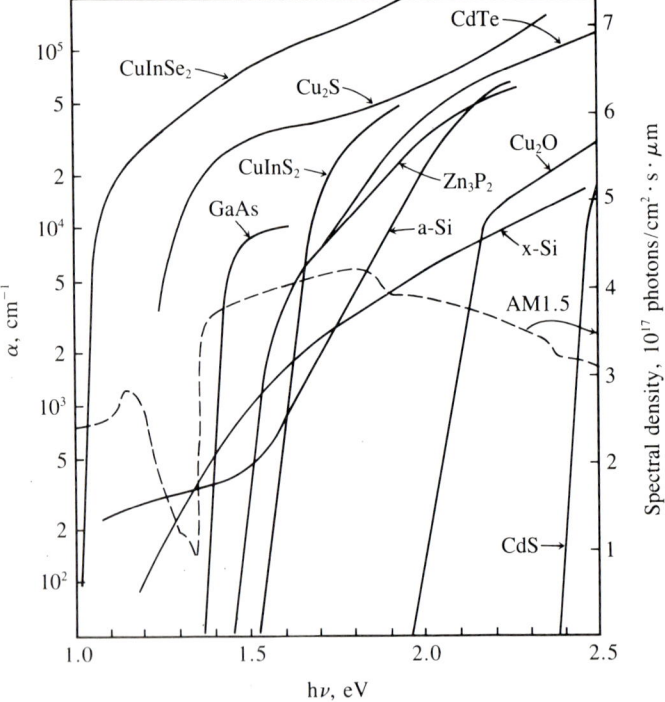

Figure 10.2 Optical absorption coefficient vs. photon energy for various semiconductors. Crystalline and amorphous silicon are denoted x–Si and a–Si, respectively here. (*Barnett and Rothwarf, 1980.* © *1980 IEEE.*)

Table 10.1 Semiconductor properties and their effects upon solar cell characteristics

Property	Characteristics affected
\multicolumn{2}{c}{Electrical}	
Size of energy gap	Portion of spectrum that is effective; V_{oc}; temperature sensitivity
Type of energy gap	Cell thickness (thickness needed is smaller if gap is direct); collection efficiency for given diffusion length
Minority-carrier lifetime	Percentage of photo-generated carriers collected; hence, J_{sc} and cell efficiency
Diffusion length	Collection efficiency; minimum allowable grain size
Conductivity types possible	Whether homojunction cell is possible
Mobility	Collection efficiency; series resistance
Possible conductivity	I^2R losses in thin front layers; junction width
Schottky-barrier height	V_{oc} and cell efficiency
Electron affinity	Shape of band diagram at Schottky and heterojunctions; collection efficiency and V_{oc}
Contact resistance	I^2R loss
\multicolumn{2}{c}{Optical}	
Absorption coefficient	Cell thickness and material required
Index of refraction	AR coating required
\multicolumn{2}{c}{Thermal}	
Thermal conductivity	Ease of cooling concentrator cell
Expansion coefficient	Epitaxy possible? Compatibility with electrical contacts and encapsulants; heat transfer
\multicolumn{2}{c}{Other physical and chemical}	
Surface properties	Carrier lifetime; ability to passivate surface
Crystal class and orientation	Chemical etching characteristics for cell manufacture and texturing; ion implantation and annealing characteristics
Lattice constant	Epitaxy possible?
Impurity diffusion along grain boundaries	Polycrystalline and cell processing feasibility owing to carrier-lifetime reduction
Chemical stability	Useful life of cell; vulnerability of cell to attack by atmosphere
Physical stability	Useful life of cell; ability to stand high-temperature processing during manufacture; possible phase changes at moderate temperatures after manufacture

Table 10.2 Properties of semiconductors*

Group(s)	Semiconductor	Energy gap at 300 K, eV	Type of gap—direct or indirect	Refractive index, n	Mobility cm²/V·s Electron	Mobility cm²/V·s Hole	Reference
			Elements				
IV	Si	1.12	ind	3.44	1350	480	1
IV	Ge	0.67	ind	4.00	3900	1900	1
VI	Se	1.74	dir	5.56 c			1
				3.72 c			
VI	Te	0.32	dir	3.07 c	1100		1
				2.68 c			
			Binary compounds				
IV-IV	SiC (α)	2.8–3.2	ind				1
IV-IV	SiC (β)	2.2	ind				1
III-V	BP	2	ind	2.6			1
III-V	AlP	2.43	ind	3.0			1
III-V	AlAs	2.16	ind				1
III-V	AlSb	1.6	ind	3.4	80		1
III-V	GaN	3.5	dir	2.4	1000	~100	
III-V	GaP	2.25	ind	3.37	50	400	1
III-V	GaAs	1.43	dir	3.4	150		1
III-V	GaSb	0.69	dir	3.9	120	120	1
III-V	InP	1.28	dir	3.37	8600	400	1
III-V	InAs	0.36	dir	3.42	4000	650	1
III-V	InSb	0.17	dir	3.75	4000	650	1
					30,000	240	
					76,000	5000 (78 K)	

II-VI	ZnO	3.2	dir		1
II-VI	ZnS (α)	3.8	dir		1
II-VI	ZnS (β)	3.6	dir		1
II-VI	ZnSe	2.58	dir		1
II-VI	ZnTe	2.28	dir		1
II-VI	CdS	2.53	dir	180	1
II-VI	CdSe	1.74	dir		1
II-VI	CdTe	1.50	dir	100	1
II-VI	HgS	2.5	dir		1
II-VI	HgSe	−0.15	dir	210	1
II-VI	HgTe	−0.15	dir	500	1
II-VI	InSe	1.3		600	2
IV-VI	PbS	0.37	dir		1
IV-VI	PbSe	0.26	dir	5500	1
IV-VI	PbTe	0.29	dir	2200	1
IV-VI	SnTe	0.18	dir		1
II-V	Zn$_3$P$_2$	1.5	dir	550	2
I-VI	Cu$_2$S	1.2	dir	1020	2
I-VI	Cu$_2$Te	1.4		1620	2
I-VI	Cu$_2$O	2	dir		2

Ternary compounds

I-III-VI	CuInSe$_2$	1.04	dir	320	2,3
I-III-VI	CuInS$_2$	1.55	dir	200	2,3
I-III-VI	CuInTe$_2$	0.96	dir	200	3
I-III-VI	CuGaSe$_2$	1.68	dir		3

Also values: 600, 930, 750 appearing in rightmost area for PbS, PbSe, PbTe; and 10, 15, 20, 20 for ternary rows.

* Data from the following sources: (1) Pankove (1971); (2) Barnett and Rothwarf (1980); (3) Chen and Mickelsen (1980). Note that estimates of index of refraction for a semiconductor of known energy gap can be obtained from the empirical rule $n^4 E_g = 77$, obeyed by semiconductors for which n^4 lies between 30 and 440.

Recently, Barnett and Rothwarf (1980) have published a unified analysis of thin-film cells, which lays out an orderly procedure for evaluating cell materials. That work will be summarized in Sec. 10.4, following discussion of amorphous semiconductors (Sec. 10.2), and miscellaneous cell materials (Sec. 10.3).

10.2 AMORPHOUS SEMICONDUCTORS

The properties of a given semiconductor, such as silicon, when it is prepared in an amorphous state, differ markedly from those of the same material in crystalline form. The energy gap and optical absorption coefficient for amorphous silicon (a–Si) are both larger than for crystalline silicon, as shown in Figs. 10.1 and 10.2, and one can readily change these properties by altering the conditions under which the materials are prepared. Thus, very thin amorphous silicon films only about 1 μm thick can be used, and one can consider economical cell designs where film properties are graded through the thickness of the semiconductor for higher efficiency.

Early a–Si films made by vacuum evaporation or sputtering in an inert gas had such high concentrations of states in the energy gap resulting from defects that the films were considered useless for solar cell application. Startling improvement occurred when up to 35 atomic percent hydrogen was incorporated in the material, forming what has been termed an *amorphous silicon-hydrogen alloy,* a–Si:H. Such a material can be prepared simply by the decomposition of silane (SiH_4) in a glow discharge, or by sputtering in a hydrogen-containing atmosphere. The hydrogen atoms bond with the unpaired electrons on Si atoms whose outermost shells of valence electrons are incomplete: in a–Si every atom does not have four nearest Si neighbors, as in crystalline silicon. With these defect states removed by the hydrogen bonding, the electrical conductivity of a–Si:H drops dramatically—a 6 atomic percent hydrogen content causes conductivity to drop by a factor of 10^7 from that of pure a–Si. Thus a–Si:H can be doped n- and p-type by the addition of atoms of phosphorus and boron, for example, by admitting phosphine (PH_3) or diborane (B_2H_6) to the glow discharge chamber along with the silane (Spear et al., 1976). Further reduction of defect state density near the Fermi level results if the active element fluorine is also present, for example by depositing from a glow discharge in SiF_4H_2, producing an amorphous alloy of silicon, fluorine, and hydrogen (a–Si:F:H).

Desirable properties of these materials are the bandgap, which is adjustable from 1.5 to 2.9 eV, the high optical absorption coefficient (Fig. 10.3), and the demonstrated ability to form n or p-type material by doping. On the other hand, free carrier diffusion lengths in the early a–Si:H were very short,

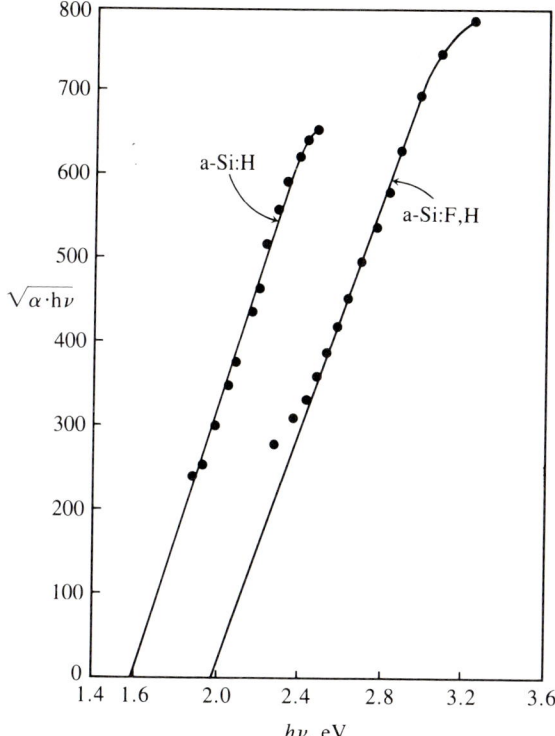

Figure 10.3 Optical absorption of amorphous silicon alloys vs. photon energy (vertical axis is square root of product of optical absorption coefficient and photon energy). (*Dalal, 1980.* © *1980 IEEE.*)

typically less than 0.1 μm, which made collection of photo-generated carriers difficult.

The report from RCA of a small-area 5.5 percent efficient AM1 a-Si:H cell (Carlson, 1977) stimulated intense and continuing interest and activity. This cell was a metal-insulator-semiconductor (MIS) device employing a thin platinum electrode on the a-Si:H. Improved material and the use of *p-i-n* structure resulted in the significant achievement of single-junction amorphous silicon cells having AM1 efficiencies greater than 10 percent, a figure often taken as the lower limit for economically viable cells for power production (Catalano et al., 1982). This cell utilizes a wide bandgap *p*-type amorphous silicon carbide window layer on underlying a-Si:H intrinsic and *n*-type layers deposited on a reflective silver back contact. By increasing material purity and reducing defects in the intrinsic region, typical diffusion lengths there were raised to 1 μm. Cell stability under illumination, a former problem, is said to be under control through attention to contamination during fabrication. Parameters for these 1.1-cm² experimental cells are: V_{oc} = 0.84 V, J_{sc} = 17.8 mA/cm², FF = 0.676, and η = 10.1 percent.

Amorphous Si cells already appear to have a place in consumer electronics, as mentioned in Chap. 7, because of their compatibility with integrated circuit fabrication techniques. The ease with which amorphous semiconductors having different energy gaps can be formed suggests that higher efficiencies may be achieved with multiple-gap amorphous cells. The multiple-gap cell, described more fully in Chap. 12, incorporates layers of semiconductors having different energy gaps; the light passes sequentially through layers having successively smaller gaps. Already an 8.5 percent efficient amorphous-silicon-based multiple-gap cell has been developed (Nakamura et al., 1982) consisting of three p-i-n structures on a supporting substrate. Light passes in turn through two different a-Si:H p-i-n cells and finally into an a-SiGe:H p-i-n cell having the smallest energy gap.

10.3 MISCELLANEOUS MATERIALS FOR SOLAR CELL USE

Several other materials deserve particular mention before we conclude this chapter with the discussion of thin-film cells (Sec. 10.4).

CuInSe₂/CdS Solar Cells

Through 1980, the only solar cells having AM1 efficiencies above 10 percent and employing semiconductors other than Si or GaAs involved epitaxial CdS, the single-crystal ternary compound $CuInSe_2$, or single-crystal InP on single-crystal CdS, discussed below. The 12 percent efficiency observed (Shay et al., 1975) with the single-crystal $CuInSe_2$/CdS has been followed by an impressive 9.4 percent AM1 efficiency in a one-square-centimeter area of polycrystalline thin-film $CuInSe_2$/CdS cell produced entirely by vacuum deposition and sputtering onto an inexpensive polycrystalline alumina substrate (Chen and Mickelsen, 1980).

Direct-gap (1.04 eV) p-type $CuInSe_2$ has a lattice constant that matches well with that of CdS, a condition favorable for epitaxy and minimizing recombination at the interface, as discussed below in Sec. 10.4. The heterojunction with n–CdS was made by successive evaporations in a single pump-down of a chamber containing Cu, In, Se, and CdS sources, onto a rf-sputtered molybdenum film on an alumina substrate. Following deposition of the selenide layers (see Fig. 10.4), which involved changing the substrate temperature in order to change resistivity, a shutter was opened to expose the

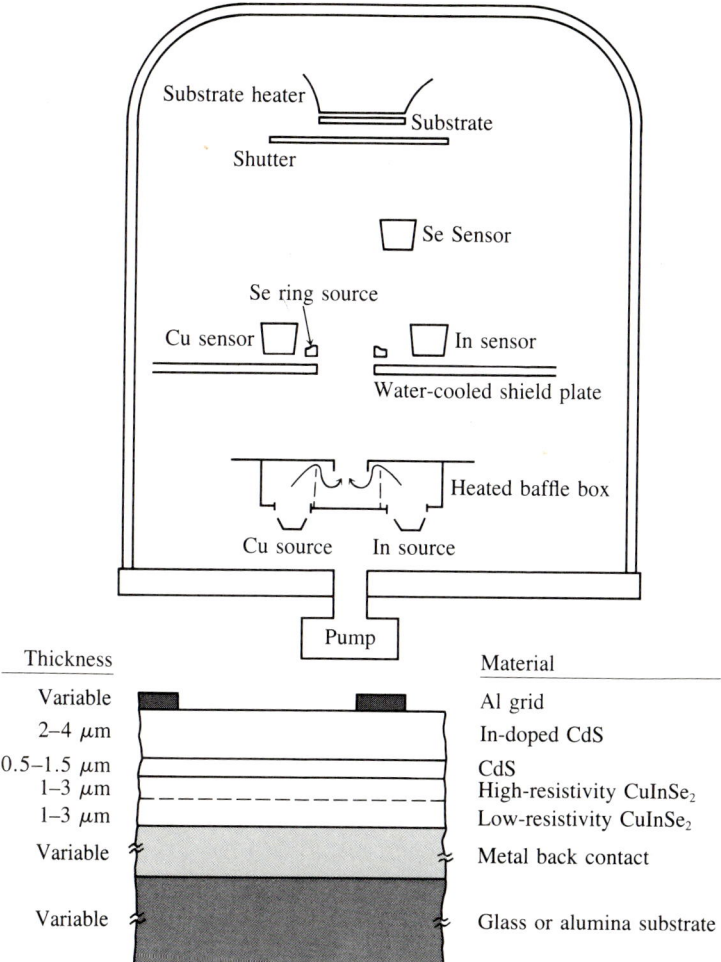

Figure 10.4 Schematic cross-section of 9.4 percent efficient AM1 CuInSe$_2$/CdS heterojunction cell formed entirely by sputtering and evaporation, as shown at top. (*After Mickelsen and Chen, 1980.*)

tantalum CdS evaporation boat which had a conventional loose quartz wool plug to prevent spattering. Finally, an aluminum contact grid was evaporated through a metal mask.

InP Solar Cells

With a nearly ideal energy gap (1.4 eV), InP has a theoretical conversion efficiency near that of GaAs, and is a candidate for use particularly in thin-film form (see Sec. 10.4). Recent achievement (Turner et al., 1980) of 15 percent AM1 efficiency in a pn-homojunction cell, formed by epitaxial growth of single-crystal p-type InP on a (100)-oriented p^+, Zn-doped InP single-crystal substrate, substantiates the expected high efficiency possible with this material. Interestingly, a native oxide made by anodic oxidation in a liquid electrolyte has been shown to produce an acceptable single-layer AR coating on this cell. Excellent cell results have also been obtained with heterojunction InP/CdS epitaxial cells (14–15 percent AM2) and indium-tin-oxide/InP heterojunction cell (15 percent AM2), indicating that InP is an outstanding solar cell material.

Tin Oxide and Indium-Tin Oxide (ITO)

Wide bandgap semiconductors may be used as transparent front contacts and AR coatings combined in solar cells. Tin oxide, SnO_2, and indium oxide, In_2O_3, having about 9 percent tin doping and commonly identified as indium-tin oxide (ITO) have bandgaps near 4 eV and bulk resistivities as low as 2×10^{-3} $\Omega \cdot$ cm, and so can function as transparent electrical conductors. They have been deposited on semiconductor substrates by spray pyrolysis, by chemical vapor deposition (CVD), and by ion beam sputtering. The oxides have indices of refraction around 2, and are typically deposited in layers about 750 Å thick if they are to be used as AR coatings, or up to several thousand angstroms thick when used as high-conductivity coatings alone. They may be used to increase conductivity of the front layer of a conventional pn-junction cell of Si, for example (Tanakura et al., 1980), or in SIS cells involving single-crystal or polycrystalline substrates, where AM1 efficiencies based on active areas of 16.2 percent and 11.2 percent, respectively, have been measured (Burk et al., 1980).

Polymer-Semiconductor Schottky-Barrier Cells $(SN)_x$/GaAs

The substance $(SN)_x$ is a metallic polymer that exhibits highly anisotropic electrical conduction. The material has a dc conductivity around 10^3 $(\Omega \cdot cm)^{-1}$ along its polymer chains and 100 times lower conductivity perpendicular to the chains. It has been shown (Cohen and Harris, 1978) that barrier heights of $(SN)_x$ on various semiconductors are higher than those of elemental metals on the same semiconductors, leading to the realization of a Schottky-barrier cell of $(SN)_x$ on GaAs (with a thin gold layer for enhanced conductivity in the front of the cell) with an open-circuit voltage as high as 0.71 V. This contrasts with the V_{oc} of gold alone on GaAs of only 0.49 V. AM1 efficiencies in excess of 6 percent without use of an AR coating were obtained in the initial test of this material combination.

Use of the polymer, which has the band structure of a highly anisotropic semimetal, appears simpler than that of the MIS approach (Sec. 11.3), which requires reliable formation of an insulating layer only 10 to 20 Å thick. To apply the $(SN)_x$ polymer, a bath of the polymer is heated in a vacuum until the material sublimes and condenses on the cooled substrate to form a layer of the desired thickness.

Organic Solar Cells

Although it is philosophically appealing to contemplate widespread use of solar cells made from thin films of organic substances such as chlorophyll, the problems seem formidable at present, and efficiencies achieved have mostly been well below 1 percent.

Organic films tend to have high but quite wavelength-dependent optical absorption coefficients, and they have high electrical resistivity, typically 10^5 to 10^8 $\Omega \cdot$ cm due to high trap densities, which lead to high series resistance, difficulty in making ohmic contacts, and space-charge limiting of current if the films are too thick.

If results to date seem disappointing it should be remembered that the efficiency with which all plants averaged together convert sunlight to biomass is only about 1 percent.

10.4 PROSPECTS FOR MASS-PRODUCED THIN-FILM CELLS

The diversity of the materials under investigation for use in thin-film cells is suggested by Table 10.3, from Barnett and Rothwarf (1980). In developing a scheme for planning material and cell development intelligently so as to reach the low-cost goals, those authors consider a generic thin-film cell structure (Fig. 10.5), divided into five functional regions;

• *Encapsulant and AR coating.* Typically the encapsulant may be glass to which the cell is attached, or a thin film of glass may be applied to the cell. The AR coating may involve layers to provide matching for transmission of light into the cell from the surrounding air, or structural alterations of the cell surface (texturing) to make the transition less abrupt.

• *Transparent contact,* which may be a conducting transparent (wide bandgap) semiconductor such as ITO or an open conducting grid—made of a material such as Ag, Ag/Ti/Pd mixtures, Au, graphite, Cu, or Ni—deposited either on the semiconductor absorber-generator layer beneath it or on the encapsulant, such as glass, if that is to be used as the supporting substrate for the entire cell.

• *Semiconductor absorber-generator* in which photons produce elec-

Table 10.3 Thin-film cell results reported for unconcentrated light

For sources of data see Barnett and Rothwarf (1980), Table 1 and References, except where noted

Absorber-generator/ collector-converter	Semiconductor thickness, 10^{-4} cm	Device area, cm^2	Absorber energy gap, eV	V_{oc}, V	J_{sc}, mA/cm^2	Fill factor	Efficiency, %	Illumination, mW/cm^2
Si/Si	25	9	1.1	.57	23.5	.72	9.5	AM1
CuInSe$_2$/CdS	1	1.2	1.04	.49	25	.54	6.6(10.6)*	100
Cu$_2$S/CdS	25	.9	1.2	.52	21.8	.71	9.15	88
CdTe/CdS	8	0.1	1.44	.75	14	.58	8.7(10.5)†	70
Cu$_2$S/ZnCdS	25	1.3	1.2	.57	19.2	.69	8.7	87
GaAs/I-M‡	25	9	1.43	.57	19.2	.60	6.5	AM1
InP/CdS	25	.25	1.3	.46	13.5	.68	5.7	74
a-Si/I-M†	1	.02	1.6	.8	12	.58	5.6	65
a-Si/a-Si	1	1.09	1.6	.84	17.8	.676	10.1§	AM1
CdSe/ZnSe/Au	2	0.01	1.7	.60	20	.45	5.0	100
CuTe/CdTe	10	6	1.4	.59	13	.63	4.8	100
CuInS$_2$/CuInS$_2$	4	.12	1.55	.41	18.9	.43	3.33	100
ZnP$_2$/Mg	10	.0025	1.5	.43	13	.45	3.0	83
InSe/Bi	10	—	1.3	.35	3	.4	1.3	70
Cu$_2$C/Cu	250	1	2	.35	7	.45	1.1	100
Merocyanine/Al	5	1	2	1.2	1.8	.25	.7	—
CdSe/ZnTe	10	.1	1.7	.56	1.89	.48	.6	85

* Efficiency reported for CuInSe$_2$/(Cd, Zn)S cell. [Mickelsen, R. A., and Chen, W. S. (1982), Record, 16th IEEE Photovoltaic Spec. Conf., to be published.]

† Efficiency reported for CdTe/CdS cell having 4-μm-thick CdTe layer on 0.1-μm-thick CdS. [Tyan, Y. S., and Perez-Albuerne, E., (1982), Record, 16th IEEE Photovoltaic Spec. Conf., to be published.]

‡ Insulator-Metal (I-M)

§ Results for p-i-n a-Si:H cell having a-SiC:H window layer. [Catalano, A., et al., (1982), Record, 16th IEEE Photovoltaic Spec. Conf., to be published.]

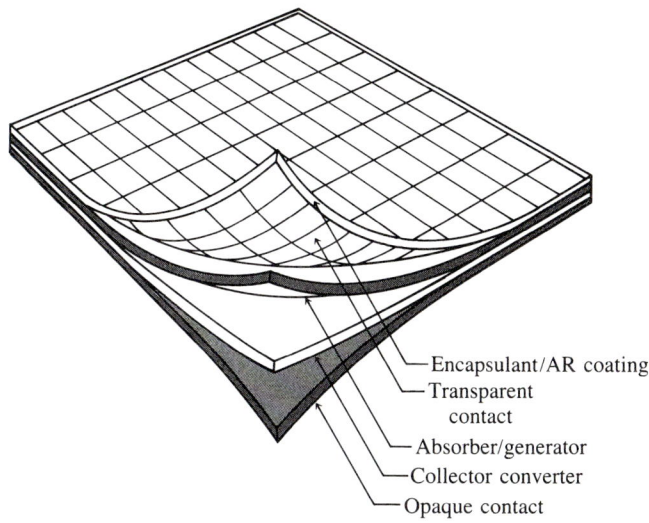

Figure 10.5 Exploded view of generic thin-film solar cell. (*Barnett and Rothwarf, 1980. © 1980 IEEE.*)

trons and holes. Requirements are a suitable energy gap and optical absorption coefficient, and a sufficiently long minority-carrier diffusion length to permit carriers to reach the collector-converter below the absorber-generator layer.

- *Semiconductor collector-converter* where minority carriers from the absorber-generator are converted to majority carriers, and which prevents the back flow of carriers. The built-in voltage between the absorber and collector regions serves as the barrier and determines how much voltage the cell can produce. This converter-collector layer must have a conductivity type opposite to that of the absorber-generator, and should have an electron affinity matching that of the absorber-generator to keep both the cell voltage and the cell current high. (*Electron affinity* is the energy required to raise an electron from the conduction band to the vacuum level, just outside the semiconductor surface.)

- *Opaque electrical contact of low resistivity.* The contact must make ohmic contact to the semiconductor collector-converter layer above it, for which it may serve as the substrate during growth of the layers of the cell. The contact is typically a thin metal such as copper or steel, and it may be required to have a high reflectivity for photons that reach it by passing through the cell without being absorbed.

Loss Mechanisms

Clearly with five distinct layers there are many possible cell materials to be considered, and it is not obvious how one should proceed in developing the thin-film cell. Following Barnett and Rothwarf (1980), we will describe the losses that in real cells can reduce efficiencies below the theoretical levels of up to 24 percent shown in Fig. 10.1. We will then summarize their results on CdS/Cu$_2$S cell development and their conclusions regarding candidates for low-cost thin-film cells.

Optical losses These are of two types: 1) reflection losses and 2) extraneous, nonproductive absorption by layers or interfaces. Optical losses may range from 5 to 25 percent, of which extraneous absorption may account for 0 to 10 percent. Reflection losses at the several interfaces run from a minimum 5 percent to as much as 18 percent, occurring at the air-encapsulant boundary (0 to 3 percent), the transparent contact (shading, 5 to 10 percent; losses in a conducting oxide if one is used, 0 to 2 percent), semiconductor interfaces (0 to 5 percent), and at the opaque contact (<2 percent).

Electrical losses These divide naturally into three categories: (1) current losses, accounting for those minority carriers not collected and converted to majority carriers; (2) voltage losses, being the reduction of V_{oc} below the maximum voltage possible due to excitation of carriers to the conduction band in the absorber-generator semiconductor; and (3) resistance losses, due to series and shunt resistance and to a poor diode characteristic, all of which reduce the fill factor at the maximum power point.

In optimized cells, current losses due to carrier recombination in the bulk may range from 5 to 20 percent, surface recombination 0 to 5 percent, and recombination at grain boundaries from none in single-crystal cells to 5 percent. Current loss due to recombination at the interface in a heterojunction cell may be substantial. If there is a lattice mismatch of amount δa there, then the interface surface-state density is about $\delta a/a^3$ cm^{-2}, where a is the mean lattice spacing for the materials that meet at the interface. This surface-state density ranges from, say, 10^{10} cm^{-2} for a very good lattice match to 10^{14} cm^{-2} for a few percent misfit. For example, in the CdS/Cu$_2$S cell one has a surface recombination velocity around 10^5 cm/s with a 4 percent lattice misfit. The net carrier loss due to the interface recombination might be less than 5 percent in optimized material, or as high as 30 to 40 percent in nonoptimized material.

Voltage losses are controlled in pn-junction and Schottky-barrier cells by the factors determining V_{oc}, namely either bulk recombination lifetime or interface and surface recombination rates, as well as the energy gap and dopant levels on opposite sides of the junction. These losses for modest doping levels may run from 4 to 50 percent. At heavy doping levels, bandgap narrowing, degeneracy effects, and Auger recombination add to the voltage losses.

Series resistance losses result from current flow over resistive paths in the front layer of the *pn*-junction cell, and so depend on the grid spacing, the thickness of the layer, and the presence of any defects that scatter carriers in that layer. Finite shunt resistance may arise from surface leakage across the junction or conducting paths produced by diffusion of dopants along grain boundaries, or possibly by tunnelling through the junction where there is an inhomogeneous doping that concentrates the electric field.

Losses due to degradation processes One would like cells to function well for 20 years, at 80 to 90 percent of their initial efficiency. Outputs may fall gradually below their initial values for several reasons. Dust and dirt accumulate, but can be removed periodically. Irreversible intrinsic degradation processes include unwanted diffusion of contact materials or dopants, especially along grain boundaries, and possibly even electromigration of material in regions where the current density is high. Extrinsic processes include oxidation of semiconductor layers if the encapsulant is breached, contact or AR coating deterioration, darkening of encapsulant due to UV exposure or weathering, and the effects of extreme temperature cycling while the cell is in use. Partial shading of a series-connected array can destroy the shaded cells, which are reverse-biased by voltage from the illuminated cells, if the cells are not provided with protective diodes.

Increasing Efficiency of Thin-Film Cells

Analysis and gradual reduction of the losses in CdS/Cu_2S thin-film cells over a five-year period led to an increase in typical laboratory efficiencies from around 5 or 6 percent to values of 9 percent by 1980. The changes made included reducing reflection from the metallic front contact by use of an evaporated contact, use of a silicon dioxide encapsulant instead of mylar and epoxy (further work is still needed on encapsulants that keep oxygen from the Cu_2S), hydrogen heat treatment to reduce recombination losses in the Cu_2S absorber-generator by improving stoichiometry, as well as heat treatments to reduce shunt resistance losses. It is expected that further development based on loss analysis will permit realizing cells having 10 percent efficiencies, with an ultimate goal of an optimized CdS/Cu_2S cell having 11.6 percent AM1 efficiency (Barnett and Rothwarf, 1980). For that cell expected characteristics are $V_{oc} \sim 0.57$ V, $J_{sc} \sim 27$ mA/cm^2, and a fill factor around 0.75.

The electron affinity match between absorber and collector can be improved by substituting 20 to 30 percent Zn for Cd in the CdS collector-converter, and so reducing the loss of open-circuit voltage occurring at the absorber-collector Cu_2S-CdS junction. An 8.7 percent efficiency has been reported for this zinc-substituted $(CdZn)S/Cu_2S$ cell, and an optimized efficiency of 15 percent is anticipated (Barnett and Rothwarf, 1980).

Cost Estimates for Production Thin-Film Cells

The cost of producing thin-film CdS/Cu_2S cells by the continuous processing line shown schematically in Fig. 10.6 has been analyzed both by methods familiar to chemical engineers and with the SAMICS methodology (Appendix 7). Results are listed in Table 10.4 for CdS/Cu_2S, and cost estimates obtained by using similar methods for many other thin-film material systems appear in Table 10.5.

The hypothetical CdS plant, scaled-up from current laboratory processing, forms the 4-μm-thick CdS or CdZnS collector-converter by vacuum evaporation on the metallic 25-μm-thick Cu:Zn alloy substrate and back contact. Copper chloride is vapor-deposited and heated to form Cu_2S, the transparent contact is printed, and the 5-μm-thick glass encapsulant is sprayed on.

The cells are assumed to average 10 percent efficiency and the plant is assumed to have a 90 percent yield. The total estimated production cost is $0.18/$W_{pk}$, of which 32 percent is material cost and 25 percent is labor cost. General and administrative costs account for about one-third, and the cost of capital for the 100-MW_{pk}/year plant amounts to about one-tenth of the final production cost. If the selling price for this large-volume item were 25 percent above the production cost, the price would be just $0.23/$W_{pk}$.

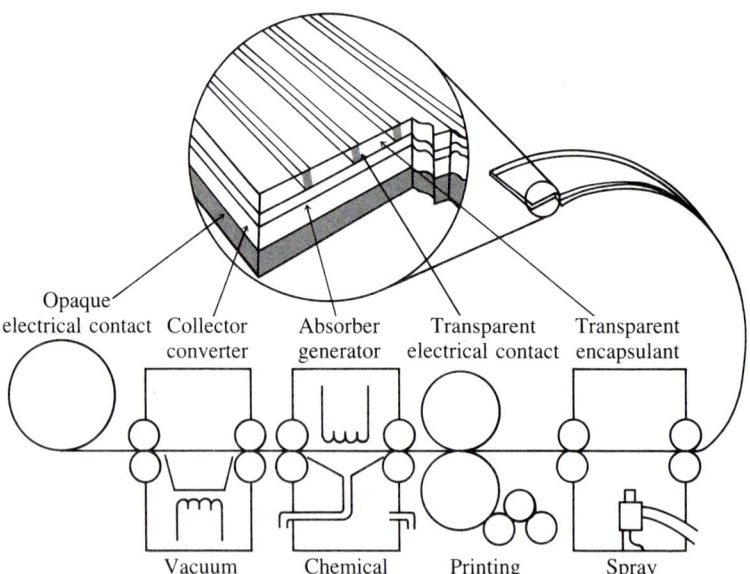

Figure 10.6 Schematic diagram of hypothetical continuous production process for making thin-film solar cells. (*Barnett and Rothwarf, 1980.* © *1980 IEEE.*)

Table 10.4 Estimated costs of producing Cu₂S/CdS thin-film cells in a continuous automated factory (in U.S. $/100 W_{pk})

See text for assumptions made. (Barnett and Rothwarf, 1980).

	Material	Production labor	Capital	General and administrative	Total
Opaque electrical contact (substrate of 25-μm-thick Cu:Zn alloy film)	3.150	.325	.065	1.520	5.060
Collector-converter (vacuum deposition of 4 μm of CdS or CdZnS)	.858	1.950	.839	1.730	5.377
Absorber-generator (vapor deposition of CuCl plus heater treatment)	.143	.955	.452	.747	2.297
Transparent electrical contact (printing of metallic grid)	1.144	.650	.194	.900	2.888
Encapsulation (spray deposition of 5-μm-thick glass compound)	.572	.650	.452	.768	2.442
TOTALS	5.867 (32.5%)	4.535 (25.1%)	2.002 (11.1%)	5.665 (31.3%)	18.064

The semiconductor materials (absorber, collector) in the cell just described contribute direct material costs of only $0.01/$W_{pk}$. The other materials whose costs are given in Table 10.5 are listed in three categories based on semiconductor material costs ($0.01/$W_{pk}$ or less, above $0.01 but less than $0.20/$W_{pk}$, and $0.20/$W_{pk}$ or more). Thickness of the absorber-generator semiconductor is assumed to be 5 μm, like that in the CdS cells, except in cases where the optical absorption coefficients permit or require thinner or thicker layers. The same assumptions of 10 percent AM1 efficiency and 90 percent yield have been made.

Although there is certainly room for discussion regarding assumptions made about efficiency and costs, one cannot help but be encouraged by the existence of so many different material systems. It appears that thin-film solar cells with manufacturing costs as low as $0.20/$W_{pk}$ may be feasible.

10.5 SUMMARY

One expects that thin films of amorphous and polycrystalline semiconductors will be used increasingly in solar cells. Amorphous silicon cells are poten-

Table 10.5 Estimated costs and selling prices of various candidate thin-film cells made in continuous production processes
Assumptions are same as for Table 10.4 except where noted. (Barnett and Rothwarf, 1980)

Material system	Thickness, μm	Material cost, $/watt	Manufacturing cost,* $/watt	Selling price,[†] $/watt
Category 1				
Cu_2S/CdS	5	.01	.181	.226
$Cu_2S/ZnCdS$	5	.01	.181	.226
a–Si/I-M	5	.01	.181	.226
Cu_2O/Cu	5	.01	.181	.226
Zn_3P_2/Mg	5	.01	.181	.226
Merocyanine/Al	.5	.01	.181	.226
Category 2				
Silicon	25	.03	.209	.261
CdTe/CdS	5	.03	.209	.261
$Cu_2Te/CdTe$	5	.03	.209	.261
CdSe/ZnTe	5	.04	.224	.280
InP/CdS	1	.06	.252	.315
Category 3				
Silicon	200	.25	.522	.653
$CuInSe_2/CdS$	5	.20	.451	.564
GaAs/I-M	5	1.00	1.587	1.984
GaAs/I-M	1	.20	.451	.564
InP/CdS	5	.30	.593	.741

*Non-semiconductor material costs are $0.167/watt for continuous process with thin-film encapsulant and opaque contact. Semiconductor material costs are increased by 42% for research, selling, and distribution charges and are assumed at 80% yield.
[†] Assumes 25% mark-up over manufacturing costs.
Note: I-M = Insulator-metal

Category 1 < $.01/watt
Category 2 < $.03 to $.06/watt
Category 3 < $.20 to $1.00/watt

tially quite inexpensive, and have demonstrated an efficiency slightly above 10 percent. Prospects appear good for achieving large-scale production of thin-film cells costing $0.30/$W_{pk}$ or less. A number of candidate materials exist, including CdS/Cu_2S, a partially zinc-substituted $CdZnS/Cu_2S$, and InP/CdS.

REFERENCES

Barnett, A. M., and Rothwarf, A. (1980), *IEEE Trans. Elect. Devices*, Vol. ED-27, No. 4, 615–630, April.
Burk, D. et al. (1980), Record, 14th IEEE Photovoltaic Spec. Conf., 1376–1383.
Carlson, D. E. (1977), *IEEE Trans. Elect. Devices,* Vol. ED-24, 449.
Catalano, A., et al. (1982), Record, 16th IEEE Photovoltaic Spec. Conf. (to be published).
Chen, W. S., and Mickelsen, R. A. (1980), *Proc. 24th Annual Tech. Symp. of Soc. Photo-Optical Instrumentation Engrs (SPIE),* San Diego, CA, 28 July–1 August (to be published).
Cohen, M. J., and Harris, J. S., Jr. (1978), *Appl. Phys. Lett.,* Vol. 33, No. 9, 812–814, 1 November.
Dalal, V. L. (1980), *IEEE Trans. Elect. Devices,* Vol. ED-27, No. 4, 662–670, April.
Mickelsen, R. M., and Chen, W. S. (1980), *Appl. Phys. Lett.,* Vol. 36, No. 5, 371–373, 1 March.
Nakamura, G., Sato, K., and Yukimoto, Y. (1982), Record, 16th IEEE Photovoltaic Spec. Conf. (to be published).
Pankove, J. I. (1971), *Optical Processes in Semiconductors,* Prentice-Hall, Englewood Cliffs, N. J.
Prince, M. B. (1955), *J. Appl. Phys.,* Vol. 26, No. 5, 534–540, May.
Shay, J. L., Wagner, S., and Casper, H. M. (1975), *Appl. Phys. Lett.,* Vol. 27, 89.
Spear, W. E., et al. (1976), *Appl. Phys. Lett.,* Vol. 28, No. 2, 105–107, 15 January.
Tanakura, H., Choe, M. S., and Hamakawa, Y. (1980), Record, 14th IEEE Photovoltaic Spec. Conf., 1186–1191.
Turner, G. W., Fan, J. C. C., and Hsieh, J. J. (1980), *Appl. Phys. Lett.,* Vol. 37, No. 4, 400–402, 15 August.

PROBLEMS

10.1 *Semiconductor properties and cell performance* Referring to Table 10.1 and Ch. 3, answer the following:

(a) As energy gap varies, how does the portion of the spectrum that is effective vary?

(b) Why does the minority carrier diffusion length affect collection efficiency?

(c) What type of semiconductor should be chosen in order to use the least amount of semiconducting material in a cell?

(d) How does carrier mobility affect the series resistance of a cell?

(e) If one uses a polycrystalline rather than a single-crystal form of a given semiconductor, which of the following properties are likely to be altered: size and type of energy gap, minority carrier lifetime, minority carrier diffusion length, conductivity types that are possible, carrier mobility?

(f) Why is minority carrier lifetime likely to be small in a direct-gap material?

(g) What constraints on cell processing does the use of a direct-gap material impose?

10.2 *Single wavelength cell* How might a photodiode designed for operation at a single photon energy, such as 1.5 eV, differ from a well-designed solar cell for use in sunlight?

10.3 *The best thin-film material* If, on behalf of a government-sponsored, long-term research program, you were to choose one or two thin-film cells for further research, which would you choose? Explain your choice and compare it (them) with the best competitors.

10.4 *Two decades from now* What are your views of the prospects, in the next 20 years, of silicon solar cells versus non-silicon cells and thin-film cells versus single crystal cells? Will they coexist or will one type dominate? Explain your reasoning.

CHAPTER

ELEVEN
VARIATIONS IN CELL STRUCTURES

CHAPTER OUTLINE

11.1 REVIEW OF SOME BASIC STRUCTURES
11.2 UNCONVENTIONAL NONCONCENTRATOR CELLS
11.3 UNCONVENTIONAL CONCENTRATOR CELLS
11.4 SUMMARY
REFERENCES
PROBLEMS

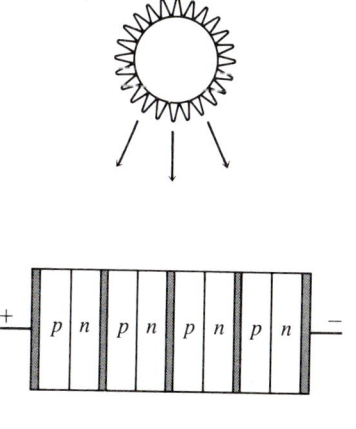

Several different cell structures have been touched upon so far in the book. These will be briefly reviewed in Sec. 11.1. The discussion will then be expanded to include other variations. We shall make no attempt to review exhaustively the large number of known novel structures nor to establish their original sources. Rather, we shall try to examine the different approaches by which to improve cell performance. Not surprisingly, the more complex cell structures have all been proposed for concentrator systems, in which high unit-area cell costs can be tolerated (see Sec. 5.2).

11.1 REVIEW OF SOME BASIC STRUCTURES

Conventional Homojunction Cell

This is the simplest structure, consisting of a *pn* junction in a single semiconductor (see Chap. 3) parallel to the surface with finger electrodes on the front surface. For materials such as Si and GaAs, in which high-quality homojunctions can be formed easily, this structure probably provides the best combination of simple fabrication and high performance.

Heterojunction Cell

Here the *pn* junction is formed by two different materials. The best example is the solar cell based on CdS, for which no reliable technology for making homojunctions exists. Instead, *n*-type CdS film is dipped in a hot $CuCl_2$ solution, or otherwise a layer of Cu_2S, which is naturally *p*-type, is deposited on the CdS film (see Fig. 7.3). Another example is the ITO (indium-tin-oxide) on Si heterojunction cell. Here the advantage is the transparency of the top semiconductor and the consequent elimination of recombination loss at the upper surface of the cell.

Schottky-Barrier (-Junction) Cell

A Schottky junction is a natural choice when *pn* junctions are difficult to make, such as in amorphous Si. The Schottky-junction structure is also popular among thin-film polycrystalline Si or GaAs cells, for which there is some fear of junction shorts due to rapid diffusion along grain boundaries. The technological challenge is to find suitable metals or surface treatments to make the reverse saturation current I_0 as low as or lower than that of *pn*-junction cells. Large I_0 translates into low open-circuit voltage V_{oc} and efficiency η. Deposition of metals thin enough for light transmission and thick enough for good electrical conductance also requires care.

11.2 UNCONVENTIONAL NONCONCENTRATOR CELLS

These cell structures were not proposed specifically for concentrator systems, although they or variations of them could also find use as concentrator cells.

Metal-Insulator-Semiconductor (MIS) Cells

As stated earlier, Schottky-junction cells tend to have high reverse saturation current, I_0. It is reasonable that I_0 could be reduced by adding an additional energy barrier—an insulator layer between metal and semiconductor. The insulator layer unfortunately can be expected to cut down the short-circuit current by hindering the carrier flow, too. The net effect is an increase in cell efficiency if the insulator is some 10 to 20 Å thick. Figure 11.1 illustrates these facts (Card, 1976).

Rather complex theories and experiments have been presented for MIS cells (Shewchun, 1980). The physical processes are still not entirely clear and the technology is under investigation.

Induced-Junction Cells

It was pointed out in Chap. 3 that it is usually desirable to reduce the thickness of the top layer of a *pn*-junction cell to avoid the recombination loss at the front surface. The ultimate shallow junction may be that of an inversion layer

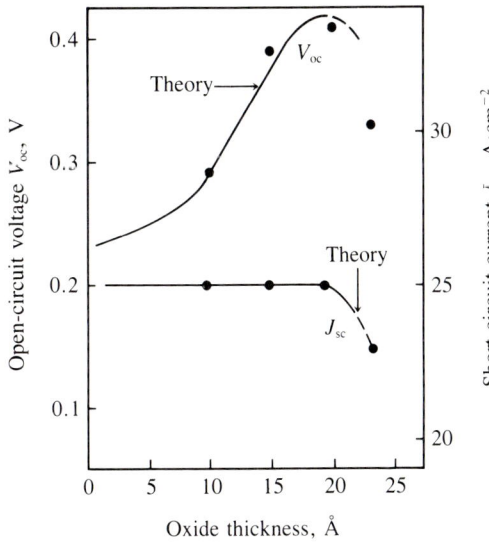

Figure 11.1 In a MIS cell, the thin insulator (oxide) between the metal and semiconductor raises V_{oc} without significantly reducing J_{sc}. Thus, adding the insulator layer improves the cell efficiency. (*Card, 1976.*)

on the surface of bulk semiconductor (Call, 1973). An n-type inversion layer may be induced by a positively biased field plate, as in an MOS field-effect transistor, or by positive charges deliberately introduced in the insulator, or both. Figure 11.2 shows an induced-junction cell with a transparent field plate. Notice that no power is consumed by biasing the field plate since no current flows through the SiO_2 insulator.

Discontinuous-Junction Cells

The pn or Schottky junction of a solar cell does not have to cover the entire cell surface. The junction may take the form of thin stripes or even a matrix of small dots and still collect carriers efficiently if the spacing between junctions is small compared to a diffusion length. For Schottky-junction cells, this is expected to decrease the reverse saturation current (smaller junction area at the same saturation current density) and hence raise efficiency (Green, 1975). For pn junctions, this will have little effect on the reverse saturation current (Hu, 1977), but may increase the short-circuit current slightly due to reduction of recombination in the diffused regions (Loferski, 1972).

Tandem-Junction Cell and Front-Surface-Field Cell

A cell can have pn junctions on both the illuminated and most of the unilluminated side (n^+pn^+ structure). Contacts are made to the p-type bulk from the back side through openings in the n^+ layer. This is called the *tandem junction cell*. In one mode of operation, the front and back n^+ regions are electrically connected. Expectedly, this structure provides a higher collection efficiency than the conventional structure, which collects carriers only at the front junction. Somewhat unexpectedly, the collection efficiency is also good when the front junction is left open. This can be explained by the fact that an

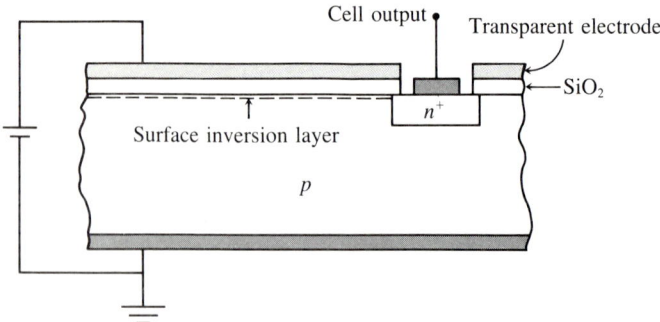

Figure 11.2 Induced junction cell. The field-induced surface inversion layer is the ultimate shallow n^+ layer. No current, therefore no power, is drawn from the biasing battery.

open-circuited n^+p junction effectively provides a low recombination velocity surface for the front side of the cell (Kim, 1980). A variation is to replace the front n^+p junction with a p^+p junction. This p^+pn^+ structure is called the *front-surface-field cell* because of its similarity, both in structure and operation, to the back-surface-field cell (Sec. 3.8). Both the open-circuited n^+p junction and the p^+p junction provide an effective low recombination surface. These cells, then, are basically the same as the interdigitated-back-contact cell discussed in Sec. 11.3.

Multiple-Pass Cells

In thin solar cells made of indirect-gap materials such as Si, it would be desirable to have sunlight make more than a single pass through the cell so as to achieve more complete light absorption. This can be done by plating the back surface with highly reflective metals (Muller, 1978). The path length can be further increased by texturing the front and/or back cell surface (see Fig. 4.8).

Liquid-Junction Cells

As an alternative to *pn* and Schottky junctions, an electrolyte/semiconductor junction can provide the built-in potential for the solar cell. The physics and chemistry of the liquid junction are less understood than for the *pn* junction, but the electrical behaviors of the liquid junction cells both in the dark and in the light are remarkably similar to those of more conventional cells. The technological challenge is to prevent corrosion of the semiconductor. Bell Laboratories have developed a cell (*Electronics*, 1981) with single-crystal indium phosphide photocathode in a solution that uses vanadium dichloride and vanadium trichloride as the ion couple in aqueous hydrochloric acid. Light shines on the *p*-type indium phosphide. Electrons diffuse to the surface, where they reduce the solution. The electrolyte, a metal anode, and the external load complete the circuit. This cell has a 11.5 percent efficiency.

One potential advantage of liquid junction cells is low cost, provided that a thin polycrystalline semiconductor is used. The electrolyte/semiconductor junction is also the building block of some photoelectrolytic cells to be discussed in Chap. 12.

11.3 UNCONVENTIONAL CONCENTRATOR CELLS

Parallel Multiple-Vertical-Junction Cell

The concept of using closely spaced multiple vertical junctions originated from the desire to maintain good collection efficiencies in satellite-mounted

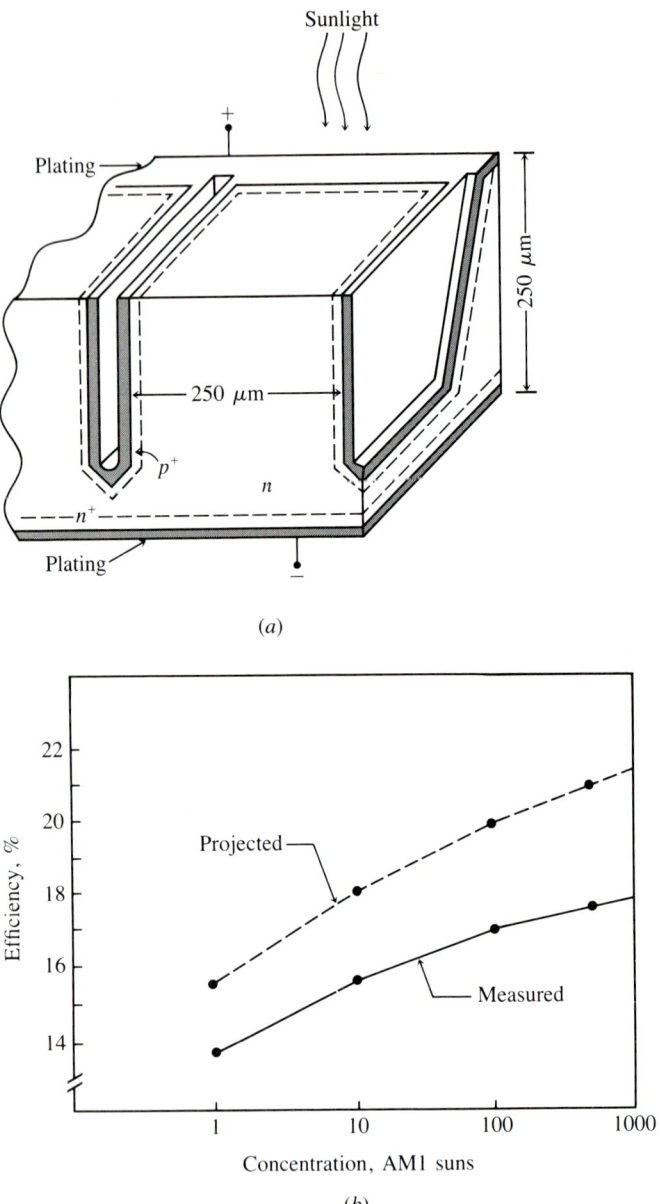

Figure 11.3 (*a*) A cell having multiple vertical junctions connected in parallel. (*b*) Measured efficiency and projected efficiency after certain improvements. Efficiency increases with the concentration ratio up to 1000 suns because of very low series resistance. (*Frank, 1980.*)

solar cells in the face of radiation-induced carrier-lifetime degradation. More recently it has been realized that some parallel multiple-vertical-junction structures, such as shown in Fig. 11.3a, can have exceedingly small series resistance (Frank, 1980). Figure 11.3b shows the measured efficiency and projected efficiency after further development on the antireflection coating and reduction of light obstruction.

Series Multiple-Vertical-Junction Cell

A concentrator Si solar cell operating under a 0.1-m^2 lens would be a power source of about 10 W at 0.6 V and 16 A. This is a somewhat inconvenient combination of low voltage and high current for interconnection or use. If the cell consisted of 16 cells in series, the cell voltage and current would be, more agreeably, 9.6 V and 1 A. A simple structure that provides the series structure is shown in Fig. 11.4. Wafers with diffused *pn* junctions are stacked and interleaved with thin aluminum foil. The stack is sintered under compression to form a solid block of serially connected *pn*-junction diodes. The block can be sliced lengthwise to yield multiple-junction solar cells (Sater, 1975). This technology is standard for producing very high voltage rectifiers. There has not been much effort in demonstrating high cell efficiency.

Series multiple-junction designs eliminate the high current and thus the series resistance limitations. One disadvantage of this and all other series multiple-junction cells is that they are sensitive to the nonuniformity of light intensity. The cell current is the smallest of the short-circuit currents of all the subcells.

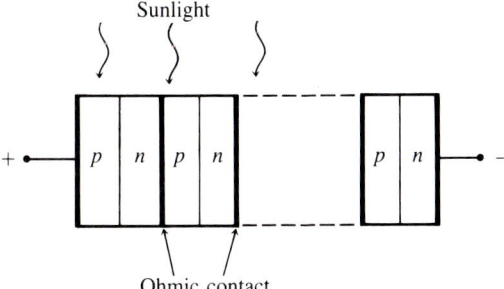

Figure 11.4 A cell having many junctions connected in series produces relatively high voltage and low current—a combination that is often desirable. However, series multiple junction cells perform well only if all junctions are illuminated at the same intensity.

V-Groove Multijunction (VGMJ) Solar Cell

This is another series multiple-junction structure. It is made with a novel manufacturing process and provided with an integral glass front cover (Chappell, 1978). Figure 11.5a illustrates the fabrication steps. The oxidized p or n Si wafer is bonded to a glass plate by a known electro-thermal bonding technique (see Sec. 9.5). Anisotropic etching, described in Sec. 4.3, is used to isolate Si into islands with V-shaped grooves. Shadowing by the neighboring islands and the oxide overhang allows n^+ and p^+ regions to be ion-implanted as well as makes possible the interconnection of the neighboring islands. Figure 11.5b shows the calculated efficiency of this cell.

Interdigitated-Back-Contact (IBC) Cell

The structure of this cell is shown in Fig. 11.6a (Schwartz, 1975). The elimination of front contacts completely removes the trade-off between contact shadowing and series resistance. In this structure, the main contribution of series resistance is from the bulk material of the cell. At high intensities, conductivity modulation makes the series resistance decrease in almost inverse proportion to the intensity. This is true for all the concentrator cells discussed in this section. The interdigitated contacts do aggravate the problem of cooling the cell to some degree. Low surface recombination velocity and long diffusion length are of critical importance to this cell and the VGMJ cell.

High-Low Junction Emitter Cell

It was pointed out in Chap. 3 that the heavily doped region of a silicon solar cell presents several problems such as very short carrier lifetime and low effective doping density. There is some experimental evidence that with the base of an n^+p solar cell doped to $0.1\ \Omega \cdot$ cm, the hole injection current into the n^+ region is 10 times the electron injection current into the p-type base. A high-low junction emitter structure has been proposed to curb the hole injection into the n^+ layer (Sah, 1978). The new structure has an n^+np configuration. The n^+ layer has the usual 0.25 μm thickness; the n layer is several microns thick. Depending on the doping concentration in the n layer, high-level injection occurs in this layer at solar concentration ratios of one to ten. When high-level injection occurs, the electron and hole densities in the

---

Figure 11.5 (a) Steps of making a V-groove multijunction cell. Step 1: grow SiO$_2$ layer, bond oxidized wafer to glass, etch V-groove pattern windows in SiO$_2$. Step 2: V-groove etch Si down to glass, ion implant n^+ and p^+ regions, anneal. Step 3: Deposit metal and alloy. (b) Calculated efficiency. (*Chappel, 1978.*)

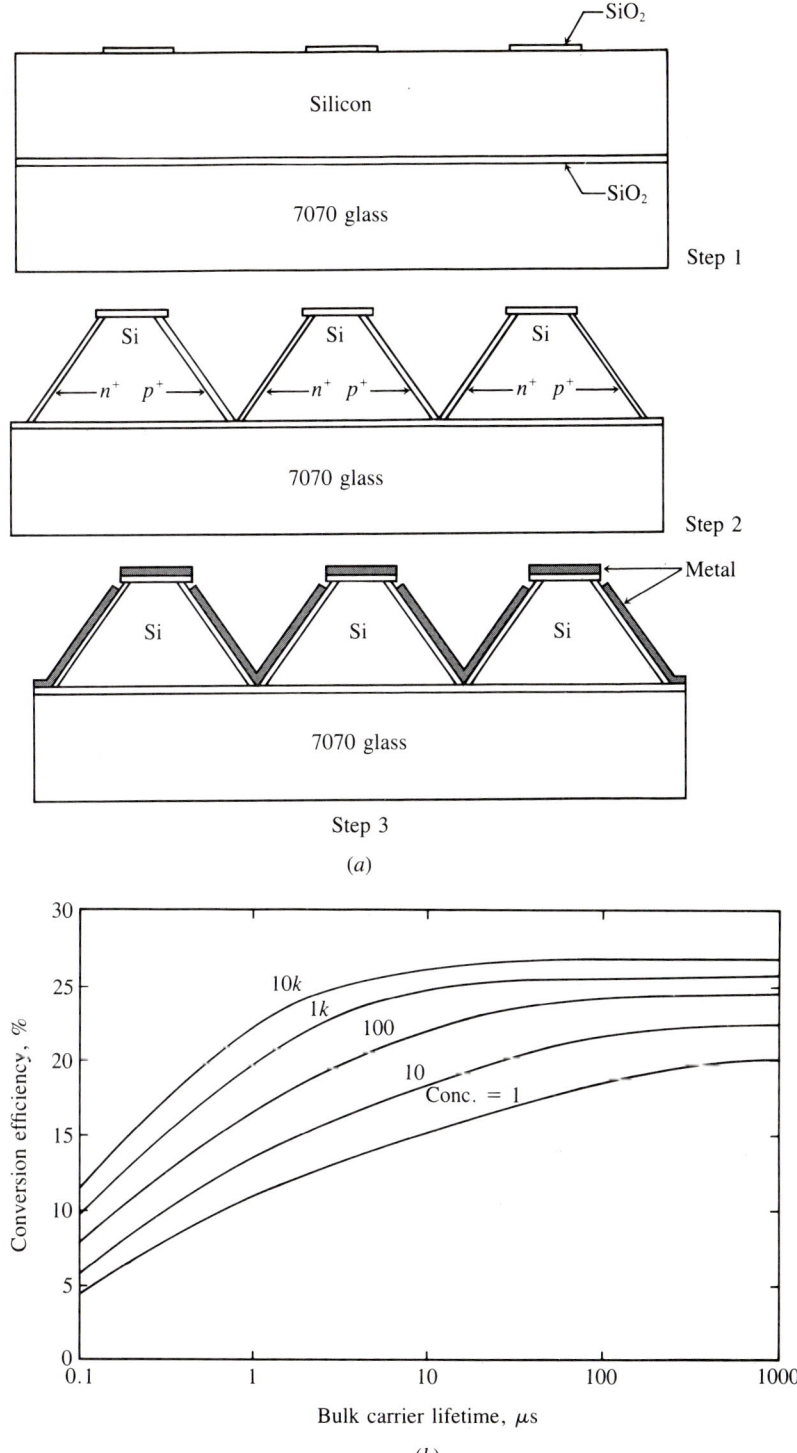

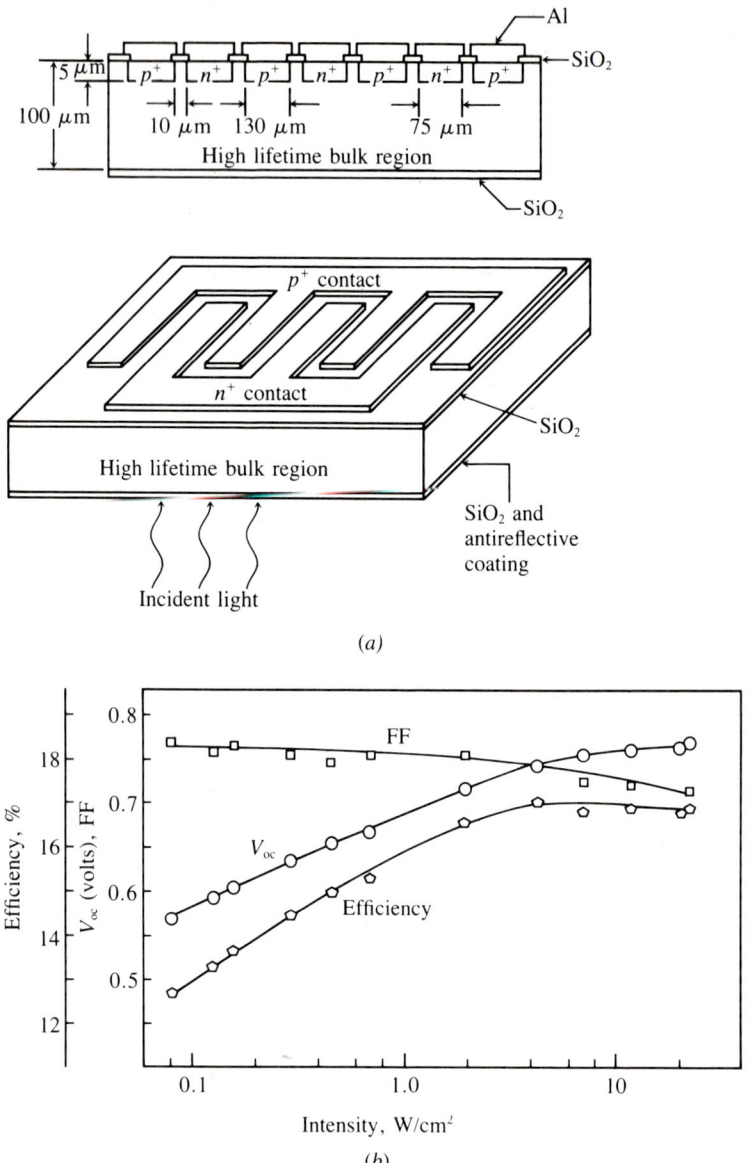

Figure 11.6 (*a*) Structure and (*b*) performance of interdigitated back contact cell. (*Schwartz, 1975*.)

Table 11.1 Calculated performance of high-low junction emitter cell (n layer is 10 μm thick doped to 1×10^{14} cm^{-3}) and conventional cell

Base resistivity is 0.1 $\Omega \cdot$ cm

	1 sun			
	J_{sc}, A/cm^2	V_{oc}, V	FF	η, %
High-low junction emitter (n^+np)	0.031	0.626	0.804	16.9
Conventional (n^+p)	0.028	0.590	0.817	14.5

	50 suns			
	J_{sc}, A/cm^2	V_{oc}, V	FF	η, %
High-low junction emitter (n^+np)	1.69	0.792	0.842	24.4
Conventional (n^+p)	1.40	0.717	0.815	17.7

n layer are about equal and electrons and holes supporting the recombination currents from the base and emitter, respectively, must both drift across the n layer. This fact fixes the emitter recombination current at about μ_p/μ_n (about 1/3 in Si) times the base recombination current. Thus the hole current injected into the n^+ region can be 30 times less than it is in the conventional structure. This results in a higher V_{oc}. Besides, a deeper pn junction can collect carriers more efficiently than a shallow junction, provided that the effective recombination velocity at the front surface is low; the n^+n junction does provide a low effective recombination velocity (see high-low junction in Sec. 3.8). The performances of a high-low junction emitter cell and a conventional cell as calculated with a simulation program (Sah, 1978) are shown in Table 11.1.

Graded-Bandgap Solar Cells

There are no physical reasons that make these cells more suitable for operation under concentrated sunlight than without concentration. Their high costs do make them less attractive for use without concentration. Some thin-film cells, however, offer the possibility of inexpensive graded-gap solar cells. The bandgap of amorphous silicon, for example, can be varied over a wide range by simply changing the film deposition conditions (Sec. 10.2).

Extending the concept of band diagram discussed in Sec. 3.2, it can be shown that a gradient in valence-band energy E_v is equivalent to an electric

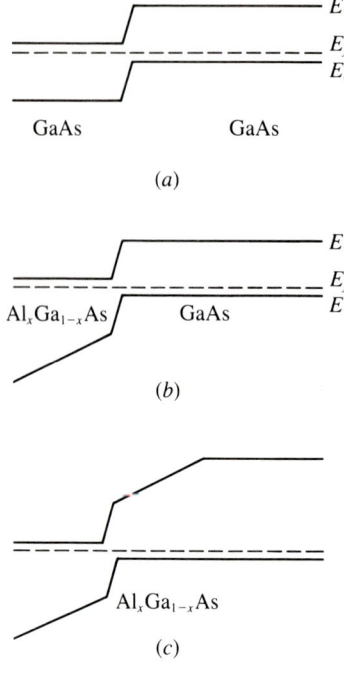

Figure 11.7 (*a*) The band diagram of the *pn* junction of a conventional GaAs cell. (*b*) The built-in drift field in a graded bandgap cell aids the collection of light-generated holes and increases J_{sc}. (*c*) When the drift field is built into both sides of the junction, the saturation current I_0 could be greatly reduced and V_{oc} increased. (*d*) This structure also can result in low I_0 and high V_{oc}.

field so far as holes are concerned. A strong drift field then can be built into a solar cell by grading the composition of a compound semiconductor and thus grading the bandgap as illustrated in Fig. 11.7*b*. AlAs and GaAs are completely miscible such that an alloy semiconductor $Al_xGa_{1-x}As$ can be formed with any *x* between 0 and 1. For $x < 0.34$, the bandgap E_g is a linear function of *x*. The resultant electric field in the graded-composition $Al_xGa_{1-x}As$ layer drives holes toward the junction and thus increases the collection efficiency. Computer analysis of the structure shown in Fig. 11.7*b* (Sutherland, 1977) predicts an efficiency of about 20 percent. This is not higher than the efficiency that can be expected from the conventional GaAs cell with proper surface passivation, as mentioned in Chap. 4 and shown in Fig. 11.7a, because the collection efficiency of the conventional cell is already very high.

Graded-bandgap structures probably can be much more rewarding if their low saturation current, I_0, is exploited. The structure in Fig. 11.7c can have small I_0 because the built-in fields work against hole injection into the n side (left-hand side) and electron injection into the p side. In Fig. 11.7d, hole injection into the n side is reduced by the field and the electron injection into the p side is naturally low because of the large E_g of this p-type material (see Chap. 3). Ideally, I_0 and hence V_{oc} of both structures shown in Fig. 11.7c and d could approach that corresponding to the wide bandgap material and I_{sc} could approach that corresponding to the smallest E_g in the structure. In this way efficiencies higher than those possible with any single-material cell might be achieved (see Fig. 3.18).

Cascade Cell

Another multi-bandgap cell that can have higher efficiency than any single-material cell, the cascade cell, will be described in Sec. 12.2.

11.4 SUMMARY

Besides the more common pn-homojunction cells, the heterojunction cells, and the Shottky-junction cells, many unconventional cell structures have been investigated in recent years; a number of them are reviewed in this chapter. Most of the structural variations described here were proposed for improving cell performance.

Many unconventional concentrator cell structures address the series resistance loss problem. These include the parallel and series connected vertical junction cells and the back-contact cells. They may be useful for operation at 100× or higher concentrations. The high-low junction emitter cell and the graded-gap cell offer possibilities of higher cell current, voltage, and efficiency.

It may be more difficult to justify the higher costs of the unconventional cell structures on the basis of improved efficiencies for nonconcentrator cells than for concentrator cells (see Chap. 5). A few unconventional nonconcentrator cell structures, however, can potentially be the basis for low-cost solar cells. For example, the semiconductor/electrolyte junction cells are potentially inexpensive. The metal-insulator-semiconductor cell structure may be more suitable for use with low-cost polycrystalline cells than the pn-junction cell structure.

REFERENCES

Call, R. L. (1973), Final Report, JPL Contract 953461.
Card, H. C., and Yang, E. S. (1976), *Appl. Phys. Lett.,* Vol. 29, 51.
Chappell, T. I. (1978), Record, 14th IEEE Photovoltaic Spec. Conf., 791.
Electronics (1981), McGraw-Hill, 46, 24 February.
Frank, R. I., Goodrich, J. L., and Kaplow, R. (1980), Record, 14th IEEE Photovoltaic Spec. Conf., 423.
Green, M. A. (1975), *Applied Phys. Lett.,* Vol. 27, 287.
Hu, C., and Edleberg, J. (1977), *Solid State Elec.,* Vol. 20, 119.
Kim, Y. S., Drowley, C. I., and Hu, C. (1980), Record, 14th IEEE Photovoltaic Spec. Conf., 596.
Loferski, J. J., et al. (1972), Record, 9th IEEE Photovoltaic Spec. Conf., 19.
Muller, J. (1978), *IEEE Trans. Elec. Dev.,* Vol. ED-25, 247.
Sah, C. T., Lindholm, F. A., and Fossum, J. G. (1978), *IEEE Trans. Elec. Dev.,* Vol. 25, 66.
Sater, B. L., and Goradia, C. (1975), Record, 12th IEEE Photovoltaic Spec. Conf., 356.
Schwartz, R. J., and Lammert, M. D. (1975), Internat'l Elec. Devices Conf., 350.
Shewchun, J., Burk, D., and Spitzer, M. B. (1980), *IEEE Trans. Elec. Dev.,* Vol. 27, 705.
Sutherland, J. E., and Hauser, J. R. (1977), *IEEE Trans. Elec. Dev.,* Vol. 24, 363.

PROBLEMS

11.1 *Classification of unconventional cells* In this chapter, the unconventional cell structures are divided, sometimes rather arbitrarily, into concentrator and nonconcentrator cells. Try, instead, to classify them as structures that (*a*) reduce costs, (*b*) reduce the series resistance effect, (*c*) improve J_{sc}, (*d*) improve V_{oc}. One structure may fall into more than one category.

11.2 *Examine a cell in depth* Select one unconventional structure described in this chapter and discuss its advantages and disadvantages in some detail.

11.3 *Invent a solar cell* You may have thought of a new variation in cell structures. (It may be a small modification to one of the structures discussed in this chapter.) Describe it and discuss its advantages and disadvantages.

CHAPTER

TWELVE
UNCONVENTIONAL CELL SYSTEMS

CHAPTER OUTLINE

12.1 MULTIPLE-CELL SYSTEMS: SPECTRUM SPLITTING AND CASCADE CELLS
12.2 THERMOPHOTOVOLTAIC (TPV) SYSTEM
12.3 PHOTOELECTROLYTIC CELL
12.4 SATELLITE POWER SYSTEMS (SPS)
12.5 SUMMARY
REFERENCES
PROBLEMS

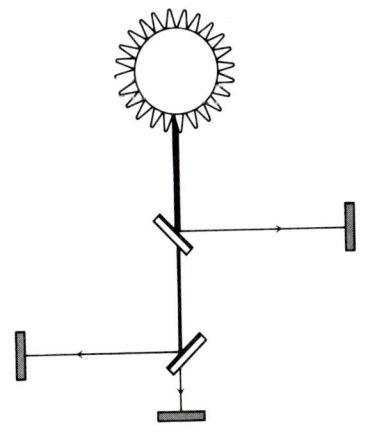

Four unconventional cell systems are described in this chapter. The first system employs a combination of individual cells, each optimized to convert the energy in a narrow spectrum of sunlight. Efficiencies higher than 30 percent are certainly within reach and 40 percent may be possible. This concept may be implemented with spectrum splitting or cascade cells.

The second system also exploits the fact that a solar cell can be very efficient in converting energy in a narrow spectrum. Called *thermophotovoltaic* conversion, sunlight is concentrated in this system to heat a refractory metal "radiator" to a very high temperature. A silicon cell then converts the blackbody radiation from the radiator into electricity. The radiation below the Si bandgap energy is reflected back to the radiator to keep it hot.

The third system does not strive for unusually high efficiency. Rather, it generates hydrogen instead of electricity as the end product. One potential advantage of such a system is that hydrogen can be stored more economically than electricity.

A fourth system puts the solar cell panel in a geosynchronous space orbit and beams power to earth via microwaves. The solar insolation in space is about five times higher than on earth and is unaffected by weather or day-night cycles. Thus this system has the unique ability to provide reliable, base-load power.

12.1 MULTIPLE-CELL SYSTEMS: SPECTRUM SPLITTING AND CASCADE CELLS

It was pointed out in Chap. 3 (Fig. 3.17) that the combined lost energy due to photon energy in excess of bandgap and energy of photons incapable of generating charge carriers account for over half of the energy entering a conventional solar cell. The loss due to excess photon energy may be reduced by increasing the bandgap E_g, but the loss due to photons incapable of generating carriers will also rise as a result. These losses would be markedly smaller if the spectrum of the light were narrower.

To take advantage of this fact, the sunlight spectrum may be split into parts with each part directed to a separate cell designed to operate efficiently for that part of the spectrum. Physically, the solar spectrum may be split with special filters, as shown in Fig. 12.1a. Each filter reflects a portion of the solar spectrum and transmits the rest.

An alternative arrangement is shown in Fig. 12.1b where cell 1 material has a wider bandgap than cell 2 material, and so on. Compared to Fig. 12.1a, the tandem cell scheme does not require the filter/mirrors but places more stringent requirements on the minimization of shadowing by the front and

12.1 MULTIPLE-CELL SYSTEMS: SPECTRUM SPLITTING AND CASCADE CELLS

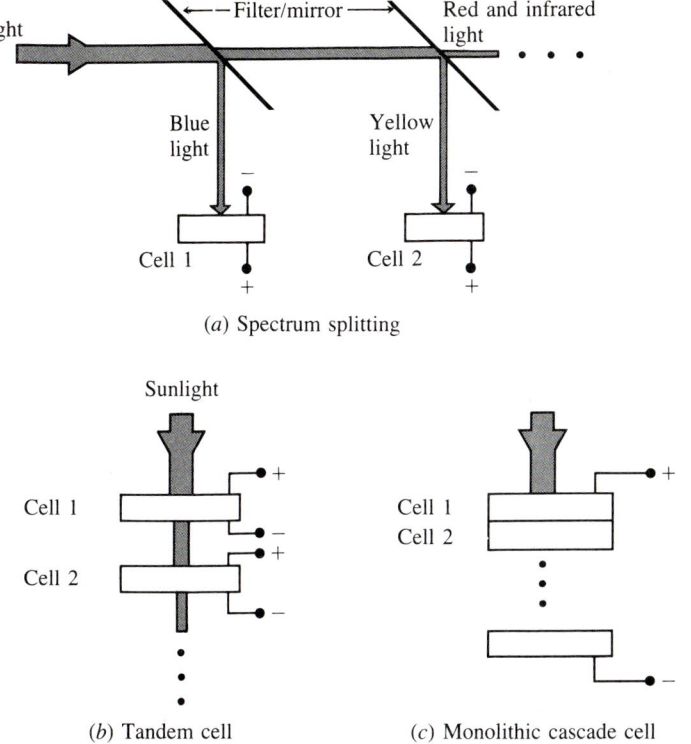

Figure 12.1 Systems employing multiple cells, each optimized for a narrow spectrum of sunlight, can have far higher efficiencies than single-cell systems. (*a*) The sunlight spectrum is split into narrow bands with wavelength-dependent filter/mirrors. (*b*) The cells double as wavelength-dependent filters. Cell 1 material has wider bandgap than cell 2 material and so on. (*c*) Similar to (*b*). In addition, cells are electrically connected in series and mechanically integrated into a single piece.

back contacts and of spurious absorption and reflection in each cell. The principle of operation, however, is the same.

Bennett (1978) has calculated the efficiencies of these two schemes and the results are shown in Fig. 12.2. The ideal system curve assumes zero contact shadowing and cell reflectance and perfect filter/mirrors. The spectrum splitting curve assumes 5 percent reflectance loss, 5 percent transmission loss at the filter/mirrors, and 5 percent shadowing and reflection losses at the cells. The tandem cell curve assumes 5 percent shadowing and reflection loss and 90 percent cell transmittance. It appears that there is little incentive to use more than two or three cells. A two-cell system should consist of cells made from 1.0- and 1.8-eV bandgap materials; and a three-cell system, 1.0-, 1.6-, and 2.2-eV bandgap materials.

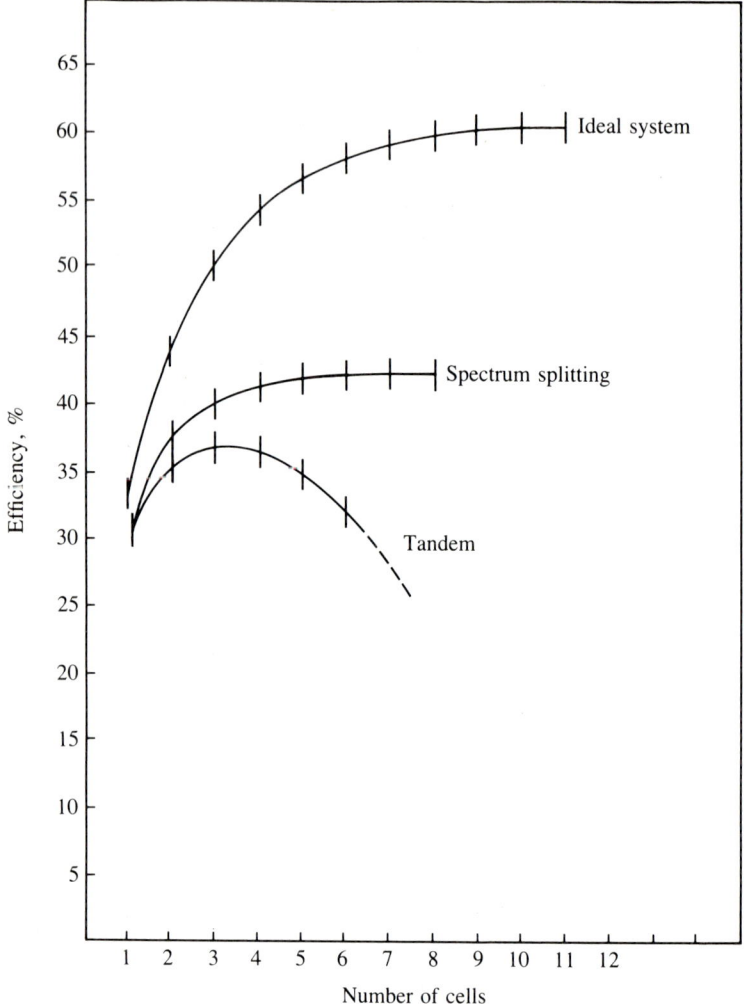

Figure 12.2 Calculated efficiency of multiple cell systems. The spectrum-splitting curve assumed 5 percent reflectance loss and 5 percent transmission loss at each filter/mirror. The tandem cell curve assumes 5 percent contact shadowing and reflection loss, and 90 percent transmittance at each cell, 1000× AM1 sun. (*Bennett, 1978.*)

A more thorough analysis of the two-cell system was made by Masden (1978), whose results are shown in Fig. 12.3. The calculated efficiency at 100× solar concentration is plotted as a function of the two bandgap energies, E_{g1} and E_{g2}. The optimum combination is 0.95 eV and 1.75 eV, which results in an efficiency of 42 percent. This is a higher efficiency than indicated in Fig.

12.1 MULTIPLE-CELL SYSTEMS: SPECTRUM SPLITTING AND CASCADE CELLS

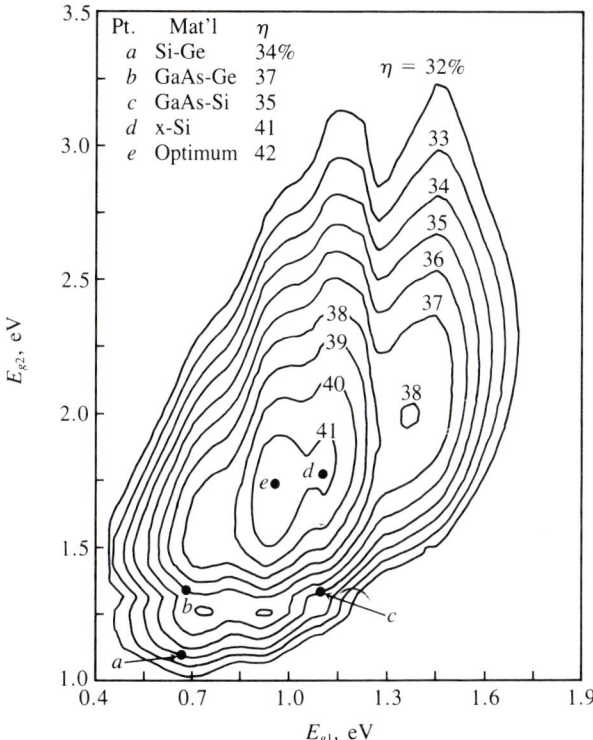

Figure 12.3 Limit conversion efficiency contours for two-cell spectrum-splitting systems for AM2 and 100× concentration. (*Masden, 1978.*)

12.2 for a two-cell system. The difference is due to the different cell efficiency models used. Several promising near-term candidates are also highlighted in Fig. 12.3. Both the GaAs/Si system and the Si/Ge system yield about 34 percent.

The measured efficiency of a Si/AlGaAs two-cell system using one filter/mirror was 28.5 percent at 165× concentration (Moon, 1978). In that experiment, the Si cell alone contributed 11.1 percent to the efficiency and the AlGaAs cell contributed 17.4 percent. The same Si cell operating under full solar spectrum yielded about 15 percent efficiency and the AlGaAs cell (1.6-eV bandgap), 20 percent. The filter/mirror used was a 17-layer dielectric filter of alternating ZnS and Na_3AlF_6.

Both the configurations shown in Fig. 12.1a and b can be operated in an alternative mode, with all cells connected in series. This mode of operation would simplify the electrical connections but would significantly reduce the system efficiency and limit the choice of cell materials. In a series connection, all cells should provide identical short-circuit currents, otherwise the output current of the system will be limited by the cell with the least current, as is the case with all series-connected cell systems.

Figure 12.1c shows a particularly attractive arrangement of series-connected cells. The design is a monolithic version of the tandem cell configuration (with internal series connection). The cells are likely made in epitaxial materials grown layer upon layer. The amount of semiconductor used is much less than in Fig. 12.1b and the transmittance of each cell is probably higher. The electrical connection between cells could be the ohmic contact between p^+ and n^+ semiconductors. For example all cells may have n^+pp^+ structures. The p^+ layer of the first and the n^+ layer of the second cell form a low-resistance ohmic junction sometimes called a *tunnel* junction. Making heavily doped layers and low-resistance junctions in wide-gap semiconductors, however, is not a simple task.

In the cascade cell, a requirement for high efficiency is that each cell produce the same short-circuit current. If one assumes an internal collection efficiency, the number of cells, and the bandgap of the bottom cell in the stack, then the bandgaps of the other cells can be determined from the requirements of equal photon flux within the spectrum absorbed by each cell. These combinations of bandgaps and the resultant efficiencies are shown in Fig. 12.4. The efficiencies are lower than those shown in Figs. 12.2 and 12.3 because the series-connected cells must satisfy the added condition of equal current from each of the sub-cells.

In designing a monolithic cascade cell one must not only consider the bandgaps of the materials in the system but also their lattice constants. High-quality epitaxial materials can only be grown on substrates of closely matched lattice constants. Figure 12.5, for example, helps determine the suitable materials for a two-cell or two-junction cascade cell. Each point represents the E_g and lattice constant of a candidate material. Each curve represents the continuous range of possibilities obtainable by mixing the materials at the two ends of the curve in varying proportions. The curve connecting GaAs and AlAs, for example, represents the E_g and lattice constant of $Al_xGa_{1-x}As$. The two bands of shaded regions represent the desired bandgaps of the two cells. These bandgaps are larger than one would deduce from Fig. 12.4, because the authors of Fig. 12.5 (Bedair et al., 1980) assumed a higher operating temperature range. The objective is to find one material from each band such that the two have closely matched lattice constants (within 0.1 Å).

Several other materials problems need consideration. The thermal expansion coefficients should be matched so that the lattice constants match at the material growth temperature as well as at room temperature. The substrate should ideally be inexpensive and readily available—Si, GaAs, Ge, InP, InAs, and GaSb in descending order of preference. Finally, the maturity of

12.1 MULTIPLE-CELL SYSTEMS: SPECTRUM SPLITTING AND CASCADE CELLS

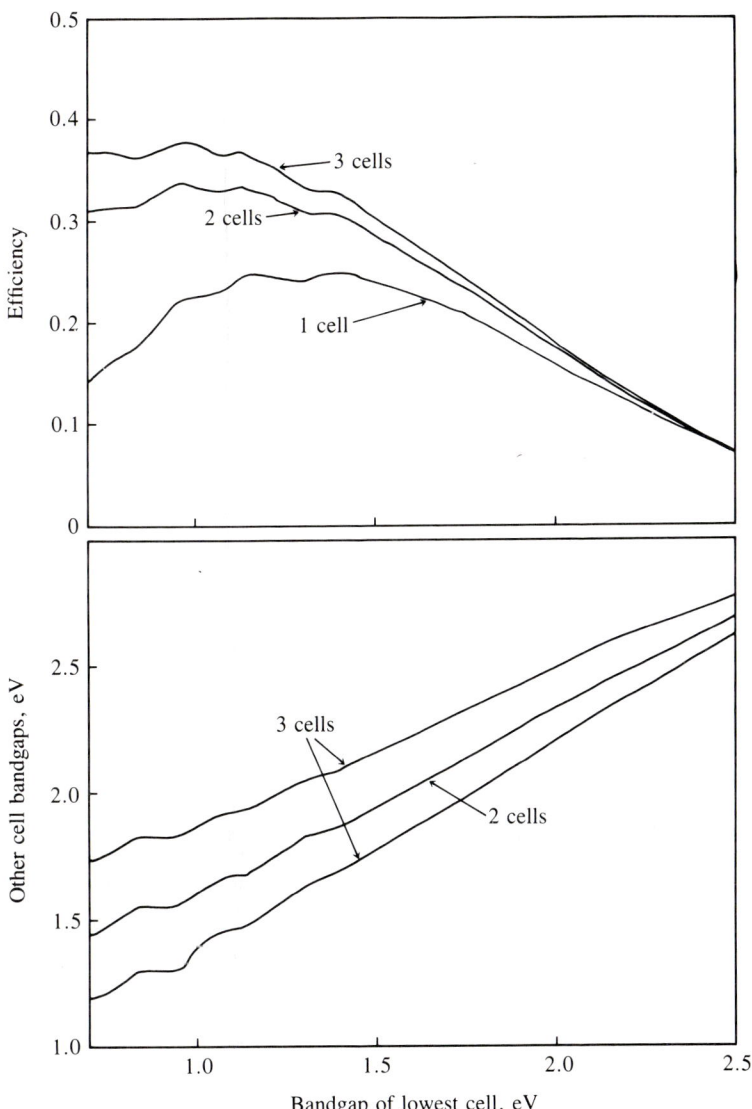

Figure 12.4 Efficiency and bandgaps of two or three cells electrically connected in series. 1000× AM2 sun, 30°C, and 80 percent collection efficiency. (*Moon, 1978.*)

the processing technology for the chosen materials is an important consideration for near-term development work. So far only experiments with GaAs/AlGaAs (Bedair et al., 1980), GaAs/GaAsP (Fraas et al., 1982), and amorphous silicon (p. 210) have been reported.

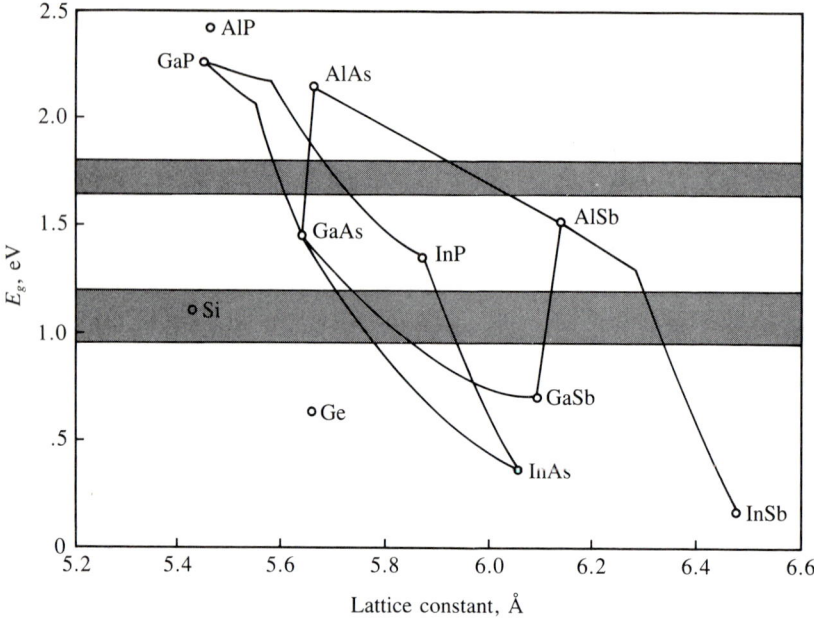

Figure 12.5 Lattice constants and bandgaps of some candidate materials for two-cell monolithic cascade cells. *(Bedair, 1980.)*

12.2 THERMOPHOTOVOLTAIC (TPV) SYSTEM

A thermophotovoltaic system generates electricity by photovoltaic conversion of radiation from an adjacent radiating surface (Wedlock, 1963). In a solar-powered thermophotovoltaic converter a parabolic mirror concentrates sunlight into a secondary concentrator (Fig. 12.6a) and finally on a refractory radiator (Swanson, 1979). The concentration ratio is several thousand so that the radiator can be heated to over 2000 K. The reason for the high temperature is that the cell efficiency is lower at lower temperatures, as shown in Fig. 12.7.

Both the unusually high efficiency and its strong temperature dependence can be explained with the diagram in Fig. 12.8. In a conventional Si cell, half of the energy is lost due to photon energy in excess of the bandgap and photons that are not absorbed. In the thermophotovoltaic system, loss of photon energy in excess of bandgap is minimized because the blackbody radiation flux decreases rapidly at energies about 1.1 eV, in accordance with Planck's law. The photons below the bandgap can be reflected from the back surface of the cell toward the radiator. If there were no parasitic absorption of these below-bandgap photons, their energies simply could not be lost. These, then, are the reasons for the high efficiency.

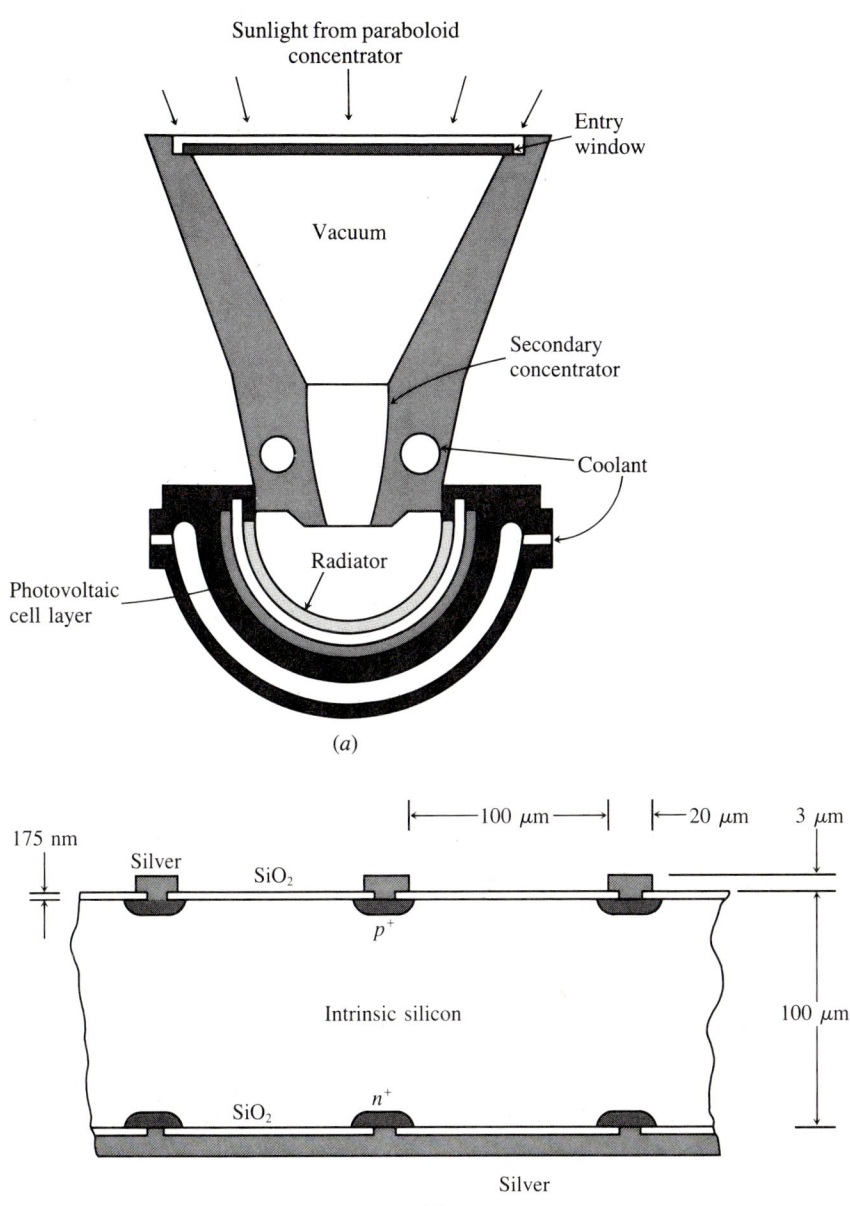

Figure 12.6 (*a*) Secondary concentrator of a thermophotovoltaic (TPV) system. (*b*) One cell structure developed for the TPV system. (*Swanson, 1979.*)

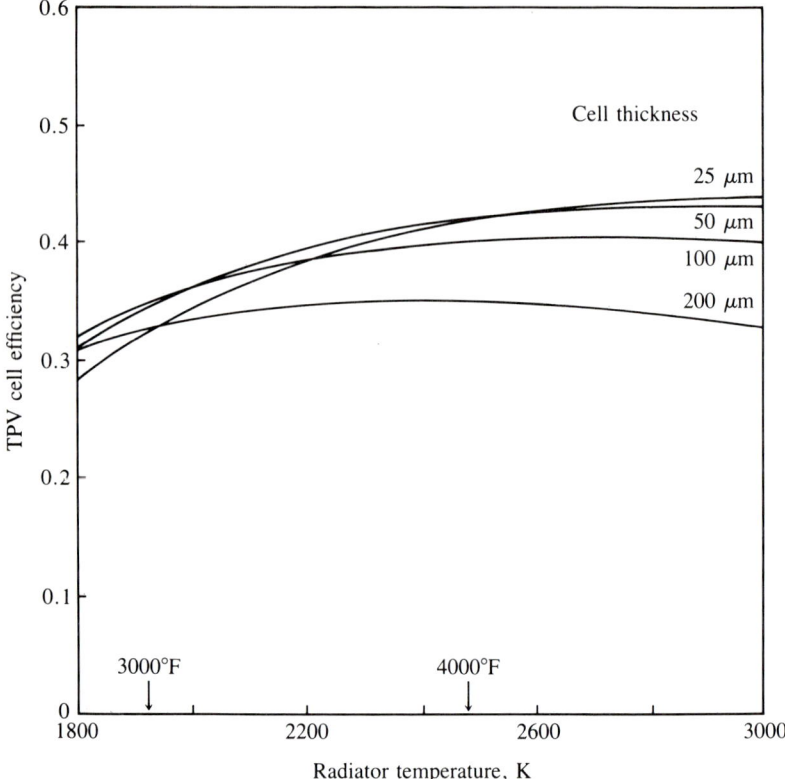

Figure 12.7 Calculated cell efficiency of a TPV cell. *(Swanson, 1979.)*

The need for high radiator temperature is explained as follows. Even at 2000 K, only 10 percent of the blackbody radiation is above the 1.1-eV photon energy. If a small percentage of the remaining 90 percent of the radiation is absorbed by the cell, the cell efficiency is significantly reduced. This becomes more serious at low radiator temperatures, at which an even smaller portion of the radiation is above the 1.1-eV photon energy.

It should be clear by now that the main task of *cell* development is to reduce the absorption of below-bandgap photons. The cell shown in Fig. 12.6b has been developed for this system. Using an electrically heated tungsten radiator, 29 percent conversion efficiency has been measured. This efficiency, although very high, is still below the projected 40 percent shown in Fig. 12.7 because of higher parasitic absorptions than assumed. There has been no report of solar energy conversion efficiency using this concept.

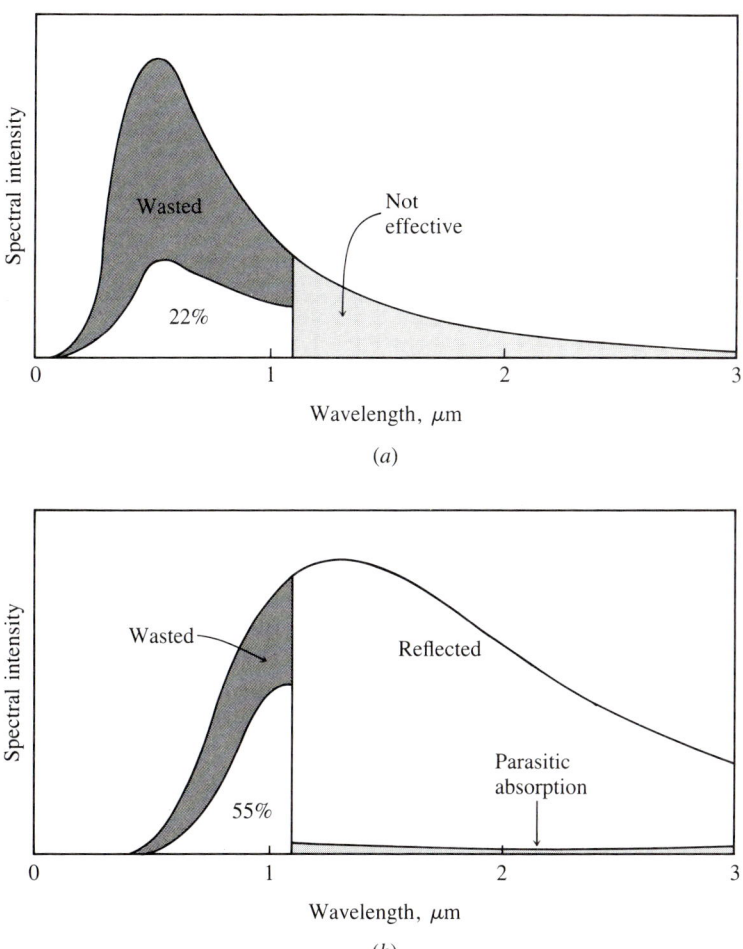

Figure 12.8 Comparison between the energy losses of (a) a conventional solar cell and (b) a TPV cell. *(Swanson, 1979.)*

12.3 PHOTOELECTROLYTIC CELL

A photoelectrolytic cell employs photovoltaic cells or photochemical processes to decompose water into H_2 and O_2. The hydrogen can be stored, burned as fuel, or used in a fuel cell to produce electrical energy. The energy conversion is from sunlight to chemical energy. The decomposition of water is the most attractive reaction, but the chemical energy may also be stored in other forms.

The photoelectrolytic cell should be distinguished from the electrochemical photovoltaic cell such as the liquid junction cell discussed in Chap. 11. An electrochemical photovoltaic cell produces electricity, and steady-state chemical reactions are to be avoided. Yet a third type of electrochemical cell is the *photogalvanic* cell, in which the incident light is absorbed by molecular species in the solution and electrical power is generated by charge transfer from excited molecular species to electrodes—a rather inefficient process. Only the photoelectrolytic cell will be discussed here.

There are two categories of photoelectrolytic cells. The first may be called the *photovoltaic-electrolysis* cell and is basically a variation of a photovoltaic cell driving a conventional electrolysis cell, as shown in Fig. 12.9a. When a sufficient voltage is applied between the electrodes of the electrolysis cell, current begins to flow and hydrogen and oxygen evolve at the cathode and the anode, respectively. The free energy change for the conversion of liquid water to H_2 demands a minimum voltage of 1.23 V. Depending on the electrode material, an "over-voltage" is also required. Platinum is the most popular electrode material because of its chemical stability and negligible over-voltage. The electrolyte also presents an ohmic drop.

Figure 12.9b shows a way of integrating the photovoltaic cell and the electrolytic cell. The photovoltaic cell may consist of two Si cells in series in order to obtain sufficient voltage. Both surfaces are coated with thin transparent Pt films and the edges are insulated from the electrolyte. Notice that no external wiring is needed.

A development (Kilby, 1979) from Texas Instruments has caused considerable excitement. It includes a potentially inexpensive way of making the photovoltaic cells used in Fig. 12.9b. A schematic of the finished photovoltaic cell is shown in Fig. 12.9c. The spheres are silicon. They are made with a process similar to that used for making lead shot. Silicon doped to 5×10^{17} cm^{-3} is melted in a tube and forced through a nozzle. Silicon droplets are formed and permitted to fall a distance of about 8 ft. During this fall the silicon solidifies. Both p-type and n-type spheres are prepared in this manner. Although some of the spheres may have internal grain boundaries, most will be suitable for solar cell application. The spheres may then be sorted by diameter into groups within 0.001-inch diameter variation.

Figure 12.9 (a) A photovoltaic cell powering an electrolysis cell. (b) The wiring in (a) is eliminated by immersing the photovoltaic cell in the electrolyte. Two solar cells are connected in series through ohmic contacts to produce the necessary voltage. (c) This cell is identical to the solar cell shown in (b) in function, but may be produced inexpensively. (d) A proposed photoelectrolysis cell system. (*Kirby, 1979.*)

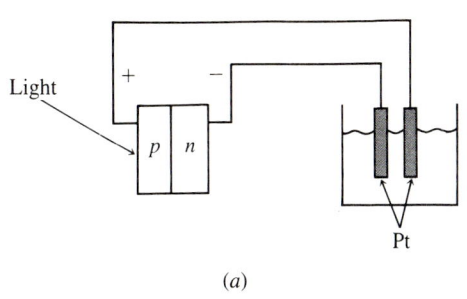

(a)

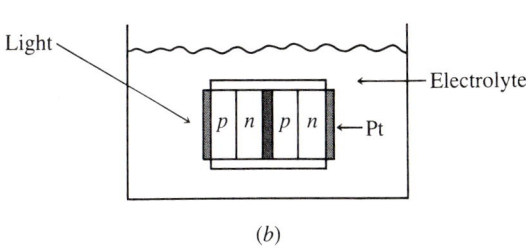

(b)

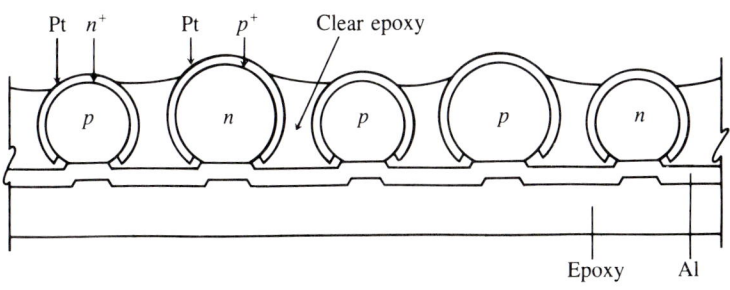

(c)

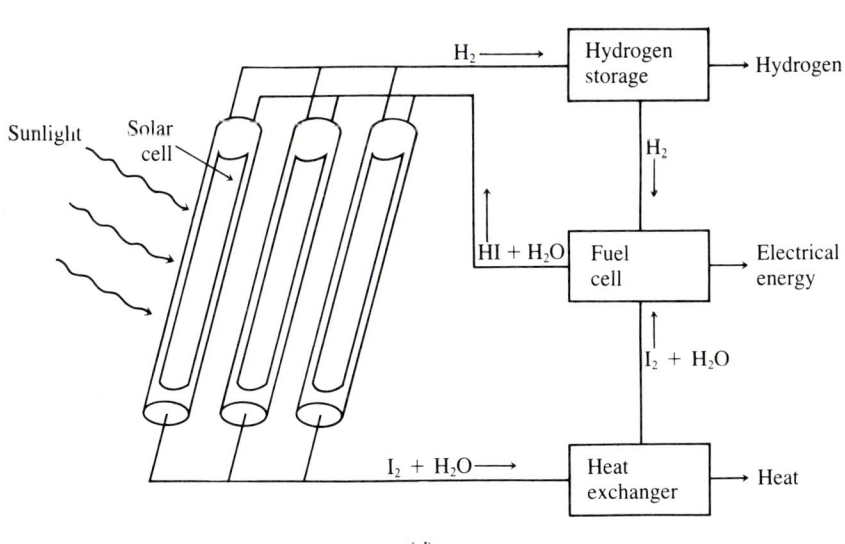

(d)

249

The p-type spheres are given an n-type surface layer and the n-type spheres are given a p-type surface layer by diffusion. A 150-Å-thick Pt film is applied to the entire surfaces of the spheres by sputtering. Platinum forms ohmic contacts to both p-type and n-type silicon. Then a 0.0005-inch layer of insulating acrylic coating is applied.

The n-type and p-type spheres are now mixed and spread on a temporary substrate coated with a thin layer of wax. The substrate is heated and the spheres are partially pressed into the wax. By inverting Fig. 12.9c and imagining the surface of the clear epoxy as the surface of the wax, one can envision the steps just described. At this time the substrate and the spheres are flooded with a layer of clear silicone resin or epoxy. After curing, the top surface of the epoxy is lapped off to expose the cores of the spheres.

The n^+ and p^+ surface layers are etched back using a concentration dependent etch. After etching, the surface of the sheet is heated, causing the feather edges of silicone resin at the top surfaces of the spheres to sag over and protect the exposed junctions. A 50-μm-thick Al layer is then plated over the entire surface and a thick epoxy layer is applied for protection and mechanical strength. The lower portion of the sheet is now washed with a solvent to remove wax and expose the Pt. Upon turning the sheet upside down, the structure shown in Fig. 12.9c is obtained. *Electrical* efficiency of 13 percent has been reported for a similar sheet (Johnson, 1981).

This is a potentially inexpensive method of making large sheets of Si solar cells. Notice that the sheet consists of many n^+p and p^+n cell pairs connected in series by the Al film. Kilby proposed to use an aqueous solution of hydrogen iodide as the electrolyte in a system shown in Fig. 12.9d. Hydrogen may be stored in a metal hydride and used to generate electricity through a fuel cell (Johnson, 1981). The total efficiency from sunlight to electricity has been about 5%. Recent work aims at developing a 32 ft^2, 300 W_{pk} module having 1500 Wh storage and an overall system efficiency of 6% (McKee, 1982).

The second category is considered by some as the *true* photoelectrolytic cell. At first sight, Fig. 12.10a and 12.10b are the same as Fig. 12.9a and 12.9b with liquid junction photovoltaic cells. Photocatalysis and reactions involving intermediate states may occur at the semiconductor surfaces in Fig. 12.10a but not the Pt surfaces in Fig. 12.9a. Nozik (1977) also pointed out that the internal potential available for driving the cell reaction may be higher than the photovoltage of the liquid junction cell measured externally.

Photoelectrolytic cells of the configuration shown in Fig. 12.10a, using a wide-gap semiconductor such as TiO_2, are almost universally used for basic studies and demonstrations. The configuration shown in Fig. 12.10b can simplify the cell. It also illustrates the possibility of using two low-energy-gap semiconductors such as Si connected by an ohmic contact to drive photoelectrolysis. The efficiency limit of photoelectrolysis using Si has been esti-

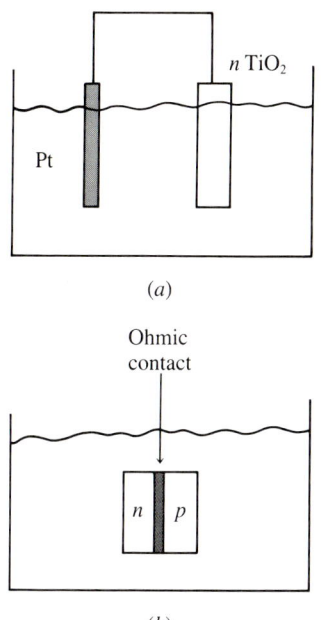

Figure 12.10 Photoelectrolysis cells in which the semiconductor/electrolyte interfaces play an active role.

mated to be about the same as Si photovoltaic cells, i.e., about 20 percent (Nozik, 1977).

12.4 SATELLITE POWER SYSTEM (SPS)

While the preceding unconventional photovoltaic systems strive to increase the energy conversion efficiency or to provide energy storage, the satellite power system would increase the solar radiation on the solar collector and eliminate the need for energy storage. In this proposed system (Figure 12.11), a satellite carrying a large solar cell panel in geosynchronous (also known as geostationary) orbit transmits power via a beam of microwaves to an antenna array on the earth's surface, where the microwaves are received and rectified into dc electricity. The advantage of placing the solar cells in space is that they will receive the full AM0 radiation more than 99 percent of the time, regardless of weather or the the day-night cycles of the earth. The solar insolation on the cells will be 32 kWh/m^2 · day. For comparison the long-term average insolation at a desert location on earth is only about 6 kWh/m^2 · day without solar tracking and 7.5 kWh/m^2 · day with solar tracking.

This imaginative system was first suggested by Dr. Peter Glaser in 1968 and has been under study by the U.S. National Aeronautics and Space Ad-

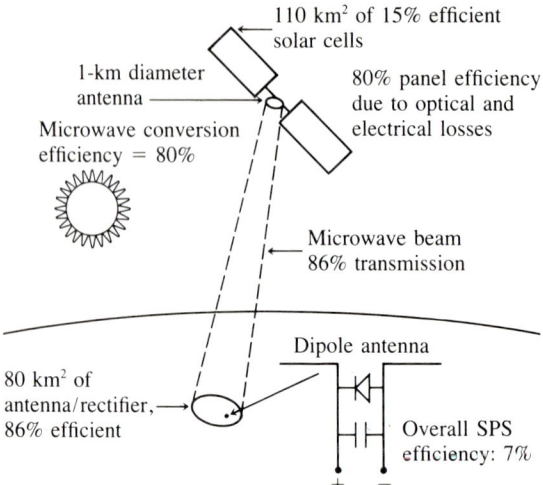

Figure 12.11 In the space power system, a satellite in geosynchronous orbit beams microwaves to earth, where the microwave is rectified into dc electricity. The numbers shown are for a proposed 10-GW system.

ministration (NASA) since 1972. Although there has been some evolution in the proposed designs of this system, the more recent conceptual designs (Brown, 1973; Glaser, 1977; DOE/NASA, 1978) differ from one another mostly in details. The following is a composite summary of these recent designs.

Each system will probably produce 10 GW (10,000 MW) of electricity on earth with an overall efficiency of about 7 percent (see Fig. 12.11). This means that the solar cell panel is about 110 km^2 in size. To save the cost of transporting materials, the SPS satellite can be built in a low earth orbit, and once completed, would itself supply power to a bank of ion thrustors. These thrustors, operating for a period of several months, would propel the satellite into a geosynchronous orbit 35,700 km above the equator. In this orbit, the satellite is stationary relative to the earth. In the earth-centered view of the sun-earth system in Fig. 2.1b, the SPS satellite would be located at a fixed point in the equatorial plane and separated from the earth by about three earth diameters. Obviously, the solar panel has to rotate once a day in order to track the sun (while the microwave antenna points at the earth all the time). The solar panel receives full AM0 radiation except when eclipsed by the earth's shadow. These eclipses occur once a day for 22 days before and after vernal and autumnal equinoxes (March 22 and September 23). These eclipses, which always occur around midnight, last no longer than 1 hour and 12 minutes, resulting in a yearly average duty cycle of over 99 percent.

Silicon solar cells operating without solar concentration at 15 percent efficiency are commonly assumed in SPS designs. The exact cell material and configuration will be determined by the usual cost and efficiency considerations and the added requirements of light weight and resistance to radiation of high-energy particles. The dc electricity is converted to 2.45-GHz (12-cm wavelength) microwaves with several million vacuum-tube-type oscillators and amplifiers. This frequency is chosen because lower frequencies would require the use of larger antennae and higher frequencies would increase the energy losses in the microwave amplifiers and raise the transmission loss in the atmosphere. The transmitting antenna has a diameter of 1 km and must generate a wavefront that is flat to one-quarter wavelength (3 cm) through tight mechanical tolerance and/or electronic control of the phases of subsegments of the antenna. The antenna must maintain a pointing accuracy measured in arc seconds and directed at a receiving antenna-array on the earth. The receiving antenna-array covers an area about 10 km in diameter and consists of a large number of dipole antenna elements, each feeding microwave power to a separate GaAs rectifier, where the power is converted efficiently into dc electricity. Because the rectifiers are small, probably several hundred million dipole elements and rectifiers will be required. The dc electricity can then be interfaced with the utility grid through an inverter.

Glaser (1977) has put the estimated system cost at $1500/kW (in 1974 dollars). According to this estimate (Fig. 12.12), the solar cell array accounts for only 25 percent of the system cost. This same array, if installed on earth, would generate 40 percent (6 kWh/m^2 · day · 15% ÷ 32 kWh/m^2 · day ÷ 7%) of the SPS output. If no credit is given to energy storage, it seems that SPS would be less cost-effective than a ground-based system, according to this rough comparison. Further reduction in solar cell cost will tip the comparison more in favor of a ground-based system. Figure 12.12 shows that the largest cost item is space transportation. More recent designs (DOE/NASA, 1978)

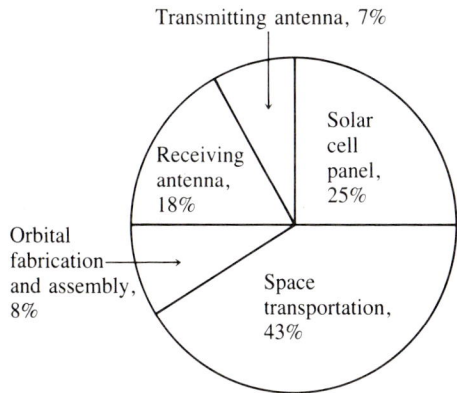

Figure 12.12 Breakdown of the estimated cost of a proposed SPS (*Glaser, 1977*). The estimated cost is $1500/kW in 1974 dollars.

placed the estimated weight of the satellite (77×10^6 kg) at over twice the estimate of Glaser (1977) making the share of the space-transportation cost even larger than that shown in Fig. 12.12. This, too, would make the comparison more favorable for a land-based system. In any event, the coming of SPS must await the successful development of low-cost space transportation technology and the technology of building large structures in space. Many people are also disturbed by the potential health hazard posed by the powerful microwave system. The intensity of the microwave beam is about 23 mW/cm^2 at the center of the receiving antenna field, gradually dropping toward the edge of the field. At the outside edge of the buffer zone surrounding the antenna field, the microwave intensity is 0.1 mW/cm^2, which is 100 times lower than the current U.S. exposure standard (10 mW/cm^2), but still ten times higher than the Soviet occupational standard of 0.01 mW/cm^2.

The Glaser (1977) article prompted about a dozen critical letters from the readers of *Physics Today*. Besides pointing to the economic, technical, and safety problems of SPS, some readers opposed the diversion of development effort from smaller, land-based photovoltaic systems. White (1977) pointed out that "Dispersed land-based converters would offer unique hope for the two million villages of the less-developed countries to maintain their individual characteristics and yet enjoy improved living conditions resulting from the availability of modest amounts of electric power obtained locally. . . ." Karr (1977) saw the size and complexity of SPS as trademarks of a military-aerospace complex and proclaimed, "I see no reason why a few unproductive aerospace corporations should be allowed to own the sun."

12.5 SUMMARY

Both the multiple-cell system and the thermophotovoltaic system avoid the inability of any single conventional solar cell to convert light in the broad solar spectrum efficiently. The multiple-cell system employs two or more solar cells each optimized to convert light in a narrow segment of the solar spectrum efficiently. The thermophotovoltaic system first shifts the spectrum to lower energies and then "recycles" the below-bandgap photons.

Both systems promise efficiencies in excess of 30 percent. They offer insights into some of the limitations of conventional solar cells and demonstrate the possibility of achieving significantly higher efficiencies than possible with the conventional cells.

The photoelectrolytic cell system eases the energy-storage problem by generating hydrogen. The photoelectrolytic cell can potentially be produced inexpensively, as demonstrated by one example.

The satellite power system has the unique advantage of providing baseload power. It is an intriguing option for the future but will probably not become economically competitive with land-based photovoltaic systems for a long time.

REFERENCES

Bedair, S. M., Phatak, S. B., and Hauser, J. R. (1980), *IEEE Trans. Elec. Dev.*, Vol. 27, 822.
Bennett, A., and Olsen, L. C. (1978), Record, 13th IEEE Photovoltaic Spec. Conf., 868.
Brown, W. C. (1973), *IEEE Spectrum,* March, 38.
DOE/NASA (1978), Reports from the Satellite Power System Concept Development and Evaluation Program, Report HCP/R-4024.
Fraas, L. M., Sawyer, D. E., Cape, J. A., and Shin, B. (1982), Record 16th IEEE Photovoltaic Spec. Conf. (to be published).
Glaser, P. E. (1977), *Physics Today,* February, 30.
Johnson, E. L. (1981), Tech. Digest of IEEE Internat'l Electron Devices Meeting, 2.
Karr, T. (1977), *Physics Today,* July, 12.
Kilby, J. S., Lathrop, J. W., Wilbur, S. C., and Porter, A. (1979), U.S. Patent 4, 136, 436.
Masden, G. W., and Backus, C. E. (1978), Record, 13th IEEE Photovoltaic Spec. Conf., 853.
McKee, W. R., Carlson, K. R., and Levine, J. P. (1982), Record 16th IEEE Photovoltaie Spec. Conf. (to be published).
Moon, R. L., et al. (1978), Record, 13th IEEE Photovoltaic Spec. Conf., 859.
Nozik, A. J. (1977), Semiconductor Liquid-Junction Solar Cells, A. Heller, ed., Electrochemical Society, Inc., Princeton, New Jersey.
Swanson, R. M. (1979), Project Report ER-1271, EPRI, Palo Alto, Calif.
Wedlock, B. D. (1963), Proc. IEEE, Vol. 51, 694.
White, R. M. (1977), *Physics Today,* July, 11.

PROBLEMS

12.1 *A simple question* Why do Figs. 12.1, 12.2, and 12.3 all assume sunlight concentration?

12.2 *Match materials* Considering only the issues of lattice constants and bandgap energy, which materials in Fig. 12.5 best meet the requirements? If Si must be one of the materials used, which material should be chosen as the second material?

12.3 *Improve the thermophotovoltaic cell* The lower illustration in Fig. 12.8 shows the spectral intensity $I(\lambda)$ of blackbody radiation. The spectral intensity of radiation from any real emitter would be $I(\lambda) \cdot \epsilon(\lambda)$, where $\epsilon(\lambda)$, the emissivity of the radiator, may be a function of λ and has values between 0 and 1. Draw a desirable ϵ vs. λ relationship, and explain your drawing.

12.4 *Efficiency and power density of TPV* The total blackbody radiation intensity is $5.67 \times 10^{-8} \cdot T^4$ W/m$^2 \cdot$ K^4, where T is the radiator temperature in degrees Kelvin. Assume $T = 2000$ K and that 10 percent of the total radiation is above the bandgap energy and 90 percent is below the bandgap energy. Also assume that the photovoltaic cell converts the above-the-bandgap photons into electricity at 50 percent efficiency and absorbs 5 percent of the below-the-bandgap radiation. What is the radiation-to-electricity conversion efficiency? What is the electric power output per square centimeter of cell area? Someone has suggested heating the radiator by burning fuel and using the fuel-powered TPV converter to drive an electric car. Would the cell/radiator area required for a 20-kW-TPV converter be practical for a passenger car?

12.5 *Another spectrum shifting idea* Certain phosphors can absorb two photons of lower energies and re-emit one photon of a higher energy. How would you make use of such phosphors to improve the efficiency of Si solar cells? What are the potential problems?

12.6 *Photoelectrolytic cell efficiency* Assume that the hydrogen produced by the system in Fig. 12.9c and d is to generate electricity through the use of a fuel cell. Assume also that the efficiencies of electrolysis and of the fuel cell are both 70 percent. What do you estimate the sunlight-to-electricity conversion efficiency of this system to be?

12.7 *Invent a process* Suggest a procedure for producing photovoltaic cells (not photoelectrolytic cells) inexpensively by borrowing ideas from the procedure for producing the cell shown in Fig. 12.9c.

APPENDIX

1
ANNOTATED BIBLIOGRAPHY

The most comprehensive source of information on photovoltaic developments is the published proceedings of the latest IEEE Photovoltaic Specialists Conference, an international conference held in the United States every eighteen months. Papers from the April 1979 West Berlin conference sponsored by the Commission of the European Communities appear in *Photovoltaic Solar Energy Conference,* R. Van Overstraeten and W. Palz, Eds. (Dordrecht, Netherlands, 1979).

Of the periodicals, *Solar Cells* (Elsevier-Sequoia, S. A., Lausanne, Switzerland) deals exclusively with photovoltaics. Some articles on PV phenomena and devices appear in applied physics and electronic device journals, such as *Journal of Applied Physics* (U.S.), *Applied Physics Letters* (U.S.), *Japanese Journal of Applied Physics, IEEE Transactions on Electron Devices* (U.S.), *Electronics Letters* (U.K.), *Journal of Physics D: Applied Physics* (Institute of Physics, U.K.), *Revue de Physique Appliqué* (France), and *physica status solidi* (a) (Akademie-Verlag, Berlin).

The *Journal of the Electrochemical Society* contains many articles on important processing technologies. The following journals on energy contain occasional articles about solar cells and their applications: *Energy Review* (International Academy, Santa Barbara, Calif.), *Energy Sources* (Crane, Russak and Co., New York), *Geliotekhnika* (Academy of Sciences, Uzbek, U.S.S.R. (English translation: *Applied Solar Energy,* Allerton Press, New York), *International Journal of Energy Research* (John Wiley and Sons, Sussex, England), *International Journal of Solar Energy* (Harwood Academic Publishers, New York), *Journal of Power Sources* (Elsevier-Sequoia, S.A., Lausanne, Switzerland), *Solar Energy* (Pergamon Press, Oxford, England, for International Solar Energy Society, Victoria, Australia), *Solar Energy Materials* (North-Holland Publishing Co., Amsterdam, Netherlands), and *Solar Engineering Magazine* (Solar Energy Industries Assoc., Washington, D.C.).

Abstract journals such as *Energy Research Abstracts* (DOE), *Electrical and Electronics Abstracts,* and *Physics Abstracts* are of course useful in searching the journal literature. Incidentally, the *Science Citation Index* permits one to find current references to specific previously published articles, and can be used to keep up to date on a particular topic.

Of the books on the subject, the 1976 IEEE reprint volume *Solar Cells*, edited by C. E. Backus, is noteworthy because of its selection of key articles published as the field developed.

Additional sources of interest are: *Application of Solar Technology to Today's Energy Needs* (Office of Technology Assessment, U.S. Government Printing Office, 1978), which surveys all solar technologies including PVs, and contains economic analyses of applications ranging from individual houses to small cities in specific locations; periodic government newsletters (such as DOE's "The Energy Consumer"); and the vast collection of reports of governmentally sponsored research and development projects (for U.S. work, contact SERI and consult the NTIS listings). Issued patents form a useful and often overlooked body of detailed and current information. Computerized literature-searching is possible in the journal abstract, NTIS report, and the U.S. patent databases.

The Solar Energy Information Data Bank, operated by SERI, is a U.S. national solar energy information network offering bibliographic and non-bibliographic database-searching capabilities; a national computer system for scientific computations, modeling, and simulation; and information dissemination services. Remote telephone access for machine searching is possible in the following databases: models, calendar of solar-related events, bibliographic entries, manufacturers' data, international projects and contracts, education, and solar legislation.

APPENDIX

2

UNITS AND RELEVANT NUMERICAL QUANTITIES

PHYSICAL CONSTANTS

Boltzmann constant: $k = 1.38066 \times 10^{-23}$ J/K

Electron charge: $q = 1.60218 \times 10^{-19}$ C

Electron mass: $m_0 = 0.91095 \times 10^{-30}$ kg

Planck constant: $h = 6.62617 \times 10^{-34}$ J·s

Speed of light in vacuum: $c = 2.99792 \times 10^8$ m/s

Thermal voltage at 300 K: $kT/q = 0.0259$ V

Wavelength of 1-eV photon: 1.23977 μm

UNITS AND CONVERSION FACTORS

Energy:

    1 kWh = 1000 Wh = 3.6×10^6 J = 860.4 kcal = 3413 Btu

    1 Btu = 2.93×10^{-4} kWh = 10^{-5} therm = 10^{-15} Q (quad)

Power:

    1 MW = 1000 kW = 10^6 W = 10^6 J/s

    1 kW = 0.239 kcal/s = 1.341 hp = 3413 Btu/h

Solar flux:

$$1 \text{ kW/m}^2 = 316.9 \text{ Btu/ft}^2\cdot\text{h} = 1.433 \text{ langley/min (cal/cm}^2\cdot\text{min)}$$

Time:

$$1 \text{ yr} = 365 \text{ days} = 8760 \text{ h} = 3.15 \times 10^7 \text{ s}$$

Length:

$$1 \text{ m} = 100 \text{ cm} = 3.281 \text{ ft} = 39.37 \text{ in}$$
$$1 \text{ mi} = 5280 \text{ ft} = 1.609 \text{ km} = 1609 \text{ m}$$

Area:

$$1 \text{ m}^2 = 10^4 \text{cm}^2 = 10.76 \text{ ft}^2 = 1549 \text{ in}^2$$
$$1 \text{ acre} = 4047 \text{ m}^2 = 0.4047 \text{ hectare}$$
$$1 \text{ mi}^2 = 2.59 \text{ km}^2 = 2.59 \times 10^6 \text{ m}^2 = 640 \text{ acre}$$

Weight:

$$1 \text{ kg} = 1000 \text{ gm} = 2.205 \text{ lb}$$
$$1 \text{ ton} = 2000 \text{ lb} = 907.2 \text{ kg} = 0.9072 \text{ metric ton}$$

SOME RELEVANT ESTIMATES

Maximum solar insolation at sea level = 1 kW/m^2

Radius of earth = 6.4×10^6 m

World solar insolation (24 hr average) = 9.2×10^{16} W = 180 W/m^2

World energy consumption rate (1978) = 9×10^{12} W

U.S. energy consumption rate (1978) = 3.6×10^{12} W

World electricity consumption rate (1978) = 7.5×10^{11} W

U.S. electricity consumption rate (1978) = 2.6×10^{11} W

Approximate energy (heat) content in

$$1 \text{ metric ton coal} = 8200 \text{ kWh}$$
$$1 \text{ barrel oil} = 1700 \text{ kWh}$$

1 gallon gasoline = 40 kWh
1 cubic foot gas = 0.3 kWh

Approximate efficiency of oil/coal electric power plant = 35 percent.

APPENDIX

3

SOLAR SPECTRUM—AIR MASS 1.5*

Wavelength, μm	$W/m^2 \cdot \mu m$	Wavelength, μm	$W/m^2 \cdot \mu m$	Wavelength, μm	$W/m^2 \cdot \mu m$
0.295	0	0.595	1262.61	0.870	843.02
0.305	1.32	0.605	1261.79	0.875	835.10
0.315	20.96	0.615	1255.43	0.8875	817.12
0.325	113.48	0.625	1240.19	0.900	807.83
0.335	182.23	0.635	1243.79	0.9075	793.87
0.345	234.43	0.645	1233.96	0.915	778.97
0.355	286.01	0.655	1188.32	0.925	217.12
0.365	355.88	0.665	1228.40	0.930	163.72
0.375	386.80	0.675	1210.08	0.940	249.12
0.385	381.78	0.685	1200.72	0.950	231.30
0.395	492.18	0.695	1181.24	0.955	255.61
0.405	751.72	0.6983	973.53	0.965	279.69
0.415	822.45	0.700	1173.31	0.975	529.64
0.425	842.26	0.710	1152.70	0.985	496.64
0.435	890.55	0.720	1133.83	1.018	585.03
0.445	1077.07	0.7277	974.30	1.082	486.20
0.455	1162.43	0.730	1110.93	1.094	448.74
0.465	1180.61	0.740	1086.44	1.098	486.72
0.475	1212.72	0.750	1070.44	1.101	500.57
0.485	1180.43	0.7621	733.08	1.128	100.86
0.495	1253.83	0.770	1036.01	1.131	116.87
0.505	1242.28	0.780	1018.42	1.137	108.68
0.515	1211.01	0.790	1003.58	1.144	155.44
0.525	1244.87	0.800	988.11	1.147	139.19
0.535	1299.51	0.8059	860.28	1.178	374.29
0.545	1275.47	0.825	932.74	1.189	383.37
0.555	1276.14	0.830	923.87	1.193	424.85
0.565	1277.74	0.835	914.95	1.222	382.57
0.575	1292.51	0.8465	407.11	1.236	383.81
0.585	1284.55	0.860	857.46	1.264	323.88

*Total energy content = 832 W/m^2.

Source: Terrestrial Photovoltaic Measurement Procedures. Report ERDA/NASA/1022-77/16, June 1977.

SOLAR SPECTRUM—AIR MASS 1.5*

Wavelength, μm	W/m² · μm	Wavelength, μm	W/m² · μm
1.276	344.11	2.388	31.93
1.288	345.69	2.415	28.10
1.314	284.24	2.453	24.96
1.335	175.28	2.494	15.82
1.384	2.42	2.537	2.59
1.432	30.06		
1.457	67.14		
1.472	59.89		
1.542	240.85		
1.572	226.14		
1.599	220.46		
1.608	211.76		
1.626	211.26		
1.644	201.85		
1.650	199.68		
1.676	180.50		
1.732	161.59		
1.782	136.65		
1.862	2.01		
1.955	39.43		
2.008	72.58		
2.014	80.01		
2.057	72.57		
2.124	70.29		
2.156	64.76		
2.201	68.29		
2.266	62.52		
2.320	57.03		
2.338	53.57		
2.356	50.01		

APPENDIX

4
ABBREVIATIONS AND ACRONYMS

AES Auger electron spectroscopy; quantitative analysis of atoms near surface of a solid obtained from energy spectrum of electrons reflected from an incident primary electron beam.

AM0, AM1, ... air mass number; number characterizing intensity of sunlight at the earth, outside the atmosphere (AM0), at surface with sun directly overhead in clear sky (AM1), etc. (see Chap. 2).

AR antireflection; refers to transparent coating or coatings of proper thickness to reduce reflection of light from front surface of solar cell.

BOS balance-of-system; elements of PV power system other than cell arrray, concentrator, and tracking systems.

BSF back-surface field; type of solar cell in which heavy doping at unilluminated back surface causes charge carriers to be repelled, thereby increasing efficiency (see Chap. 3).

CIS conductor-insulator-semiconductor; composite solar cell structure having conductor separated by an insulating layer from semiconductor (see Chap. 11), similar to MIS.

CPC compound parabolic concentrator; nonimaging concentrator for solar cells employing parabolic reflecting surfaces set at an angle to each other (see Chap. 5).

CVD chemical vapor deposition; growth of solid layer on substrate in heated chamber supplied with gas containing atoms that form the layer.

CZ Czochralski; growth of large single-crystal ingot by slowly withdrawing a seed crystal in contact with molten material (see Chap. 4).

DLTS deep-level transient spectroscopy; determination of energy levels of trapping centers from, usually, the temperature dependence of the rate of decay of the capacitance of a *pn* or MOS capacitor made on the semiconductor under study, following a step change in the bias voltage across the capacitor.

ABBREVIATIONS AND ACRONYMS 265

DOE Department of Energy (U.S.); responsible for energy research and development.
EFG edge-defined film-fed growth; growth of (nearly) single-crystal ribbons, typically of silicon, for solar cells by solidification through a die (see Chap. 9).
EPI epitaxial (Gr.: *epi* = on, upon; *taxis* = arrangement); refers to growth of a crystalline film that nucleates with the correct orientation because of the spatial periodicity of the substrate.
ERDA Energy Research and Development Administration (U.S.); predecessor to U.S. Department of Energy.
FF fill factor; factor in solar cell efficiency, equal to current-voltage product at maximum efficiency divided by short-circuit current times open-circuit voltage (see Chap. 3).
FZ float zone; purification of semiconductor by heating so that a molten zone of the material passes through an ingot, taking impurities with it.
IBC interdigitated back contact; solar cell having electrodes on back instead of front surface to eliminate reflection or absorption of incident light.
IC integrated circuit; miniaturized solid-state electronic circuit.
IEE Institution of Electrical Engineers (U.K.); professional society.
IEEE Institute of Electrical and Electronics Engineers (U.S.); professional society.
ITO indium-tin oxide; large bandgap semiconductor employed as optically transparent electrode for conventional or Schottky-barrier solar cells.
LDC less-developed countries; nations other than the Western and the centrally managed economies, typically having relatively low per-capita incomes.
LED light-emitting diode; semiconducting *pn* junction, which emits light when electric current passes through it.
LPE liquid-phase epitaxy; growth of oriented crystalline solid film from a liquid in contact with an underlying substrate in a heated chamber.
MBE molecular-beam epitaxy; growth of solids layer-by-layer from beams of impinging atoms in an evacuated chamber.
MIS metal-insulator-semiconductor; similar to MOS, having metal and semiconductor separated by an insulator.
MOS metal-oxide-semiconductor; type of semiconductor device employing metal electrode separated by an oxide insulator from an underlying semiconductor.
NAA neutron activation analysis; quantitative determination of constituent atoms in a solid from their characteristic gamma-ray emissions following neutron bombardment.
NASA National Aeronautics and Space Administration (U.S.); responsible for U.S. space program.

NSF National Science Foundation (U.S.); responsible for U.S. support of basic research and development, primarily in universities.

NTIS National Technical Information Service (U.S.); distributes reports on U.S. government-sponsored research and development projects.

PEC photoelectrochemical; solar energy conversion involving use of liquid electrolyte and production of electricity and/or fuel gases (see Chap. 12).

PR photoresist; photon-sensitive emulsion used to delineate areas to be selectively etched away during fabrication of semiconductor devices.

PV photovoltaic; abbreviation used as in "PV system."

RBS Rutherford back-scattering; quantitative determination of atoms at or near the surface of a solid from the characteristic energy loss suffered in the back-scattering of a beam of incident alpha particles.

RTR ribbon-to-ribbon; crystallizing or enlarging grains in polycrystalline silicon ribbon stock by heating ribbon as it moves between supporting rollers (see Chap. 9).

SEM scanning-electron microscope; electron microscope that forms images of objects from secondary electrons produced by a scanned primary elctron beam.

SERI Solar Energy Research Institute (U.S.); organization in DOE responsible for solar energy research and development.

SHAC solar heating and cooling of buildings.

SIMS secondary-ion mass spectroscopy; quantitative determination of composition near surface of solid through mass spectroscopy of atoms sputtered away by impact of incident beam of ions such as argon.

SIS semiconductor-insulator-semiconductor; composite structure having typically one wide- and one narrow-bandgap semiconductor separated by insulator.

SOC silicon-on-ceramic; polycrystalline film obtained by cooling a ceramic sheet wiped with or dipped into molten silicon (see Chap. 9).

SPE solid-phase epitaxy; regrowth without melting of an amorphous layer upon underlying crystalline substrate, as in scanned laser annealing of semiconductors.

SPS satellite power system; proposed system having orbiting solar panels producing electricity beamed to earth on microwave beam for rectification and use (See Chap. 12).

SSMS spark source mass spectroscopy; quantitative determination of composition of a solid by mass spectroscopy of ions produced when sample is vaporized in a spark between inert electrodes.

STEP solar thermal electric power generation.

TPV thermophotovoltaic; process of producing electricity with solar cell illuminated by blackbody radiation from radiator heated by solar or thermal energy (see Chap. 12).

VGMJ V-groove multijunction; multiple-junction solar cell, for use with concentrator, produced in part by V-groove anisotropic etching (see Chap. 11).

VMJ vertical multijunction; multiple-junction solar cell, for use with concentrator, having series-connected current-collecting junctions oriented parallel to direction of sunlight incidence (see Chap. 11).

VPE vapor-phase epitaxy; growth of solid films, upon an underlying solid substrate, from a vapor in a heated chamber.

APPENDIX

5

TABULATION OF DEMONSTRATED CELL EFFICIENCIES BY CELL TYPE

Material	V_{oc}, V	J_{sc}, mA/cm^2	FF	η, %
Silicon	0.622	34.3	0.796	16.8
Silicon	0.731	1380	0.745	18.3
Silicon	0.783	15000	0.75	17.6
Silicon	0.621	36.5	0.806	18.3
Silicon	0.540	32.7	0.76	13.3
Silicon	0.585	31.9	0.74	13.8
Silicon	0.572	24.2	0.76	10.5
Silicon	0.522	28.1	0.79	11.5
Silicon	0.561	26.2	0.778	11.4
Silicon	0.50	15	0.63	5
Amorphous-Si	0.84	17.8	0.676	10.1
Amorphous-Si	0.880	13.1	0.57	6.6
Amorphous-Si	0.878	11.1	0.66	6.4
GaAs	0.97	25	0.81	20
GaAs	0.93	28	0.81	18.2
GaAs	1.05	270	0.85	23
GaAs	—	—	—	21
GaAs	0.95	23	0.78	17
GaAs	0.76	24.4	0.63	12
AlGaAs/GaAs	2.05	10.8	0.74	16.5
GaAs	0.56	22.7	0.67	8.5
Cu$_2$S/CdS	0.52	24.8	0.71	9.2
Cu$_2$S/ZnCdS	0.6	22.8	0.75	10.2
CuInSe$_2$/CdS	0.4	38	0.63	9.4
CuInSe$_2$/CdZnS	0.431	39	0.631	10.6
CdS/CdTe	—	—	—	10.5
CdS/CdTe	0.79	—	—	8
CdS/InP	0.63	15	0.71	12.5
InP	0.66	24.8	0.64	11.5
WSe$_2$	0.72	22.6	0.57	10.2
MoSe$_2$	0.65	25	0.56	9.4
CdTe	0.723	18.7	0.64	8.6
CdSe	0.57	23.8	0.48	6.5
CdSe/ZnSe/Au	0.6	20	0.45	5.0
CuTe/CdTe	0.59	13	0.63	4.8
Zn$_3$P$_2$	0.48	18	0.55	4.3
CuInS/CuInS$_2$	0.41	19	0.43	3.3
Merocyanine/Al	1.2	1.8	0.25	0.7

* For additional listings see Table 9.1 (on rapid silicon-growth techniques) and Table 10.3 (on thin-film cells).

Area, cm²	Remarks	Ref. no.
2	p^+-n-n^+	1
2	41× concentration	1
0.4	500× concentration	2
2.8	MIS, active area	3
2.8	semicrystalline, MIS	3
4	edge-supported growth	4
5	silicon-on-ceramic	
11.4	ITO/semicrystalline	4
45	EFG (ribbon)	5
0.04	roller quenching	6
1.1	a-SiC/a-Si p-i-n	7
0.73	MIS	4
1.2	p-i-n	4
0.5	n^+-p-p^+ homojunction	4
1.5	AlGaAs/GaAs	8
1.5	AlGaAs/GaAs, 10×	8
—	AlGaAs/GaAs, 1000×	9
0.5	CLEFT	4
0.1	on Ge-coated Si	4
4.0	tandem cell	4
9	CVD thin film, MIS	4
1	solution ion-exchange	4
1	16% Cd, 84% Zn	4
1	evaporation	4
—	20% Cd, 80% Zn	10
—	thin film	10
—	electrodeposition	10
0.23	single crystal InP	11
—	PEC	4
—	PEC	4
—	PEC	4
0.02	electrodeposition, Au/CdTe	4
1	electrodeposition	4
0.01		12
6		12
1	Mg/Zn$_3$P$_2$, thin film	4
0.12		12
1		12

REFERENCES

1. Fossum, J. G., Nasby, R. D., Burgess, E. L. (1978), Conf. Record, 13th IEEE Photovoltaic Spec. Conf., 1294.
2. Frank, R. I., and Goodrich, J. L. (1980), Conf. Record, 14th IEEE Photovoltaic Spec. Conf., 423.
3. Green, M. A., et al. (1980), Conf. Record, 14th IEEE Photovoltaic Spec. Conf., 684.
4. Feucht, D. L. (1981), Conf. Record, 15th IEEE Photovoltaic Spec. Conf., 648.
5. Kalejs, J. P., et al. (1980), Conf. Record, 14th IEEE Photovoltaic Spec. Conf., 13.
6. Arai, K. I., et al. (1980), Conf. Record, 14th IEEE Photovoltaic Spec. Conf., 31.
7. Catalano, A., et al. (1982), Conf. Record, 16th IEEE Photovoltaic Spec. Conf., to be published.
8. James, L. W., Moon, R. L. (1982), *Appl. Phys. Lett.*, Vol. 26, 467.
9. Kaminar, N., et al. (1982), Conf. Record, 16th IEEE Photovoltaic Spec. Conf., to be published.
10. Conf. Record, 16th IEEE Photovoltaic Spec. Conf. (1982), to be published.
11. Bachmann, K. J., and Buehler, E. (1975), *Appl. Phys. Lett.*, Vol. 26, 229.
12. Barnett, A. M., and Rothwarf, A. (1980), *IEEE Trans. Electron Devices*, 615.

APPENDIX

6
SOLAR CELL EXPERIMENTS

Of greatest interest to the user of a cell is the maximum power P_{max} that the cell can deliver to a resistive load under given conditions of illumination. Related to that are the maximum cell efficiency η_{max} and the load resistance $(R_L)_{max}$ for maximum power output. Also of interest are the short-circuit current I_{sc}, the open-circuit voltage V_{oc}, and the fill factor FF. One will want to observe the variation of output power with time during the day as the sun moves across the sky, and possibly also the effect of ambient temperature on cell output. The reader interested in a deeper understanding of solar cell operation may also want to verify the dependence of I_{sc} and V_{oc} upon intensity of illumination. Several experiments for measuring these charactistics are outlined here.

EQUIPMENT REQUIRED

For Experiment 1 you require only the following simple apparatus:

1. Solar cell. Single cells may be purchased from hobby shops or the suppliers listed in Appendix 8.
2. Digital or analog dc voltmeter and milliammeter, or multimeter. The internal resistance of the voltmeter should be at least 1000 Ω/V to permit measurement of open-circuit voltage.
3. Assorted resistors in the 1 to 150 Ω range.
4. Light source. The sun or a lamp may be used.
5. Hookup wire.

In addition to this equipment, used in measuring I_{sc}, V_{oc}, the entire I-V characteristic, cell output power, and the optimum load for maximum power output, the simple curve tracer of Experiment 2 requires only a low-frequency oscilloscope having horizontal and vertical channels and a step-down transformer to reduce the 110- or 250-V line voltage to a few volts. For Experiment 3, on the dependence of I_{sc} upon intensity of illumination, you will require a light source such as a flashlight or a microscope illuminator that provides a fairly uniform, slightly diverging beam. A photographic exposure meter is useful but not necessary.

EXPERIMENT 1

Connect the solar cell and meter as shown in Fig. A6.1. The cell may be illuminated with sunlight or by artificial light from a lamp. With no load resistor connected and the voltmeter (or multimeter set on "dc volts") connected as shown, measure V_{oc}. With an ordinary silicon solar cell the value should be around 0.6 volts. Next connect the dc milliammeter (or multimeter set on "dc milliamperes") across the cell and read I_{sc}. The value will depend on the area of the cell and the intensity of illumination. For a typical 2 cm by 2 cm silicon cell in AM1 sunlight, I_{sc} should be around 120 mA. Multiply V_{oc} by I_{sc} to determine the maximum possible power the cell could provide if it had a "square" I-V characteristic.

To determine the output under different load conditions, connect different resistors R_L across the cell as shown. Values should range from 1 to 150 Ω. Measure and tabulate for each value of R_L the voltage across points A–B, and the current obtained by putting the milliammeter (or multimeter set on "dc milliamperes") in series with the load resistor R_L. Compute the power output $P = I \times V$ for each load resistance for which it occurs. The ratio of the maximum power to the total input power, given by the product of the intensity of illumination and the area of the cell, is an approximate value of the cell efficiency. (For a typical silicon cell this will be at most about 15 percent.)

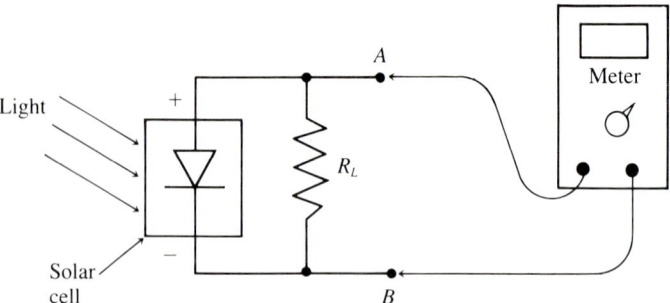

Figure A6.1 Set-up for Experiment 1.

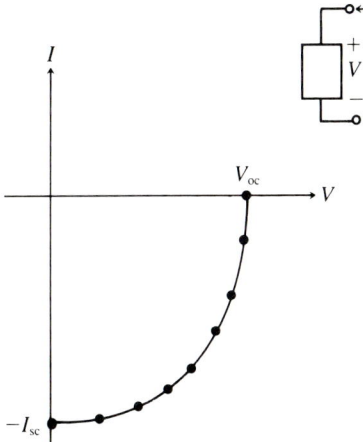

Figure A6.2 Typical plot of I vs. V for solar cell, taken with different load resistances R_L connected to cell. Note current and voltage conventions shown in inset.

For a more accurate value of the maximum efficiency, plot the pairs of values I and V you measure, draw a smooth curve through them (Fig. A6.2), and find the maximum power point. Note that with this method you will only obtain points in the fourth quadrant of the I-V characteristic, where the cell is delivering power to the load. The optimum load resistance for the cell will be $(R_L)_{max} = V_{max}/I_{max}$, where V_{max} and I_{max} are the voltage and current associated with the maximum power point. From the true maximum power, determine the maximum cell efficiency. Calculate the fill factor, FF $= P_{max}/I_{sc}V_{oc}$, which should be 0.7 or greater and is a measure of how "square" the I-V characteristic is.

EXPERIMENT 2

By using an oscilloscope having both horizontal and vertical input channels, you may display the I-V characteristics of solar cells quickly.

Connect the circuit shown in Fig. A6.3, using a step-down transformer to reduce the ac voltage from the mains to a few volts. Set the horizontal sensitivity to a volt or so per division, and display the voltage applied to the solar cell on the horizontal axis. The vertical deflection is proportional to the voltage across the series resistor R_S. If your oscilloscope has a vertical sensitivity of 10 mV per division, a 50-Ω resistor R_S will yield a 10-division vertical deflection if a 2-mA cell current flows. With the circuit connected as shown, the cell voltage will be displayed with positive values to the right, and the current with negative values upward. This simple curve tracer will display the entire I-V characteristic and the fourth quadrant will appear in the upper

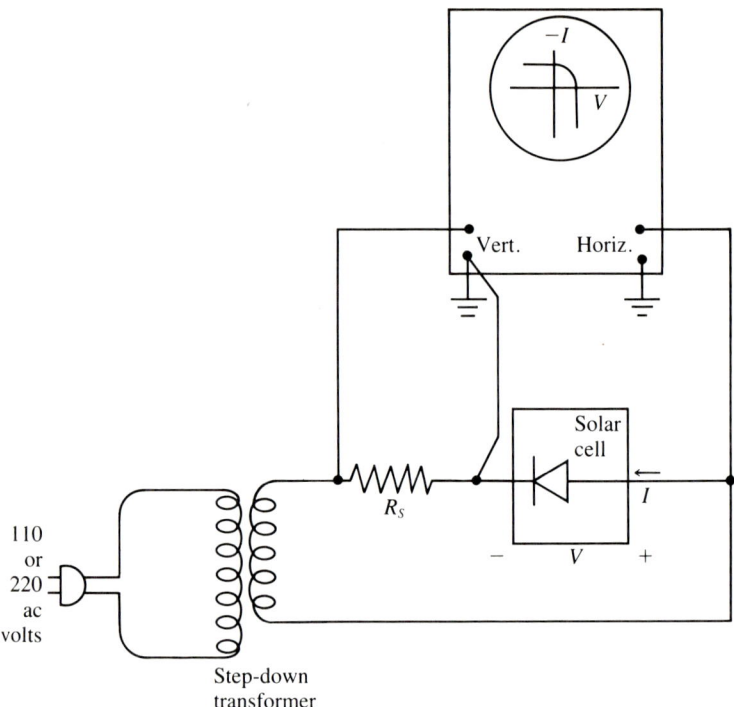

Figure A6.3 Circuit for using oscilloscope to display I-V characteristic of solar cell.

right corner of the screen. If you have a transistor curve tracer you can, of course, use it to display the I-V characteristics.

Connect the circuit and calibrate the display, finding the origin of the plot (all voltages off) and the vertical calbration (current = deflection in divisions × deflection sensitivity in volts per division ÷ R_S). Observe the I-V characteristic of the cell in the dark (this is the characteristic for a conventional pn-diode rectifier) and then gradually increase the illumination and note how the characteristic moves to increasingly negative currents as the photon-generated current from the cell increases.

EXPERIMENT 3

Using either the meters of Experiment 1 or the curve tracer of Experiment 2, you may observe the variation of the cell output with intensity of illumination.

Measure the short-circuit current each hour through the day with the cell illuminated by sunlight and oriented perpendicular to the sun at noon. Plot and interpret the results, considering that to the first order, I_{sc} is proportional to the intensity of illumination and that the intensity reaching the interior of the cell decreases as the sun goes lower in the sky (air-mass number increasing) and as the angle of incidence of light on the cell increases.

If you have a way of measuring the intensity (see Chap. 2), you can calibrate your solar cell absolutely. Even without such equipment you may verify that I_{sc} is proportional to intensity of illumination, as follows.

Place a light source such as a flashlight or a microscope illuminator that projects a uniformly bright, slightly diverging beam, on a stable support. If you have a photographic exposure meter use it to determine two distances from the light at which the intensity differs by a factor of 2 (they are the distances at which the exposure meter readings differ by one f-stop or by a factor of 2 in shutter speed). Measure I_{sc} at each location and compare the values. If you do not have an exposure meter, you may use the following technique.

On a piece of cardboard draw two concentric circles whose *areas* are in the ratio of 1:2. Your solar cell should just fit inside the smaller circle. With the cardboard find the distance from the light source at which the beam just fills the larger circle, then place the solar cell there and measure I_{sc}. Now find with the cardboard the location closer to the light source where the beam just fills the smaller circle, and place the solar cell there and again measure I_{sc}. The intensity in the second location is approximately twice that in the first, and so the short-circuit current in the second case should be about twice that in the former.

OTHER EXPERIMENTS

You may see the deleterious effect of increasing temperature on cell performance by observing I_{sc} and V_{oc}, or the entire I-V characteristic as you *gently* heat the solar cell by bringing a soldering iron *near* its back surface. If you have a pulse generator and several light-emitting diodes (LEDs) you may be able to determine the effective minority-carrier lifetime in the cell by illuminating the cell briefly with light from the LEDs and observing the decay of I_{sc} after the light pulse ceases (values of a few to tens of microseconds are typically obtained with silicon cells). If you have colored filters with known transmission characteristics you can place them between the cell and the sun or a bright white light, such as an indoor movie light, and verify that the cell will not respond to very long wavelength radiation in the infrared portion of the spectrum. With a monochromator you can actually measure the wavelength at which response ceases and obtain an approximate value of the energy gap of the semiconductor used in the cell.

APPENDIX

7
COMPUTER SIMULATIONS OF PHOTOVOLTAIC CELLS AND SYSTEMS

A solar cell can be represented very roughly by a simple equivalent circuit containing a current source, a diode, and resistors, and the sun can be modeled as a blackbody at 5743 K. In a slightly more precise representation, such an equivalent circuit could be derived for each wavelength and the outputs of the circuits summed, each being weighted according to the relative intensity at that wavelength in the actual solar spectrum. But for serious exploratory solar cell studies and PV system development, such simple approaches are inadequate, and more comprehensive models and computer simulations are required. Three different types of computer program and data input are used:

Simulations of solar cell operation. These programs model mathematically the optical and electrical phenomena occurring in cells. When the cell dimensions and impurity distributions are provided along with solar spectral data, the typical program calculates the *I-V* characteristics, the efficiency and its components such as fill factor, and perhaps other quantities such as the excess carrier concentration at each point in the cell and the variation of cell output with temperature.

Economic and complete PV system simulations. Programs provide information ranging from the manufacturing costs for cells producing a given peak power under standard insolation to such quantities as average cost and annual energy production for cells at a particular location used with specified equipment for concentration, tracking, power conditioning, and energy storage.

Databases and special analysis programs. Examples are solar insolation data and electrical circuit and heat transfer analysis programs.

We shall discuss one particular solar cell simulation program, PERSPEC, as an example, and then comment on some other programs in the three categories. Incidentally, some programs may be available for distribution, but the reader is warned that even if programs have been written in a high-level language such as Fortran, they usually require some modification to run properly on a different computer, because of minor differences in compilers, operating systems, and subroutine libraries.

SIMULATIONS OF SOLAR CELL OPERATION

An example of a cell simulation program is the one-dimensional Fortran program PERSPEC, written by T. I. Chappell (1978). It combines multilayer optical simulation with semiconductor analysis to make a fast and accurate simulation program for photovoltaic solar energy converters. The program uses about 16,000 bytes of central computer memory with memory overlays, and requires about 2 seconds of execution time on a CDC-6600 computer to find the short-circuit current, open-circuit voltage, and the maximum-power point on the I-V characteristic of a typical silicon p-i-n photovoltaic device in AM1 sunlight. About half a minute is required for a device in 200X sunlight. The program has been used successfully to simulate several different types of solar cells and a special sensor for very intense sunlight (Chappell, 1976).

Table A7.1 lists features of this large program, which was based on the work of Gwyn, Scharfetter, and Wirth (1967) and Scharfetter and Gummel (1969). The user inputs the cell dimensions and doping profiles, and chooses the other items listed under item I in the table. The doping profiles may be obtained from analytic expressions, or from a process program such as SUPREM, which will output the impurity profile resulting from the use of any of several standard silicon integrated circuit fabrication procedures, such as oxidation or ion implantation. (For information about SUPREM, contact the Technology Licensing Office, Stanford University, Stanford, Calif.)

The PERSPEC program contains as data (item II of the table) the solar spectrum, and electrical and optical characteristics of silicon, such as carrier mobilities and diffusion constants, lifetime, as well as the real and imaginary parts of the indices of refraction of silicon and various AR materials. Equations or tables in the program represent the variations of these characteristics with temperature, local electric field, doping, and excess carrier concentrations.

The program first solves the optical problem (item III) to find the light intensity throughout the cell and, from that, the rate of optical generation of mobile carriers at each point. The entire cell is divided into about one hundred layers, with thinner layers near cell surfaces or interfaces where electric fields and carrier densities change most rapidly. The program solves numerically

Table A7.1 Features of the PERSPEC solar cell simulation program

I. User supplies these quantities:
  Cell thickness
  Cell doping profiles (spatial distribution of impurities)
  Description of antireflection coatings employed
  Metal backing on cell
  Assumed cell temperature
  Sunlight concentration factor

II. Program contains these data and represents these phenomena:
  AM1 solar spectrum
  Carrier mobilities in silicon, and their dependences on doping, excess carrier concentrations, electric field, and temperature
  Carrier lifetime and its dependence on doping; Auger recombination
  Real and imaginary parts of indices of refraction of silicon and various antireflection coating materials

III. Program solves these problems and equations:
  Reflection, transmission, and absorption of photons
  Spatial distribution of optically generated mobile carriers
  Nonlinear semiconductor carrier-transport equations
  Poisson's equation

IV. Printed or plotted output:
  Conversion efficiency, short-circuit current, open-circuit voltage, fill factor
  I and V at maximum power output
  Illuminated and dark I-V characteristics
  Spectral response
  Accounting for all optical and electrical energy losses
  Spatial distributions of mobile carriers and electric fields

the semiconductor carrier transport equations and Poisson's equation, relating change of electric field to local charge concentration, starting from a first approximation for electric fields based on the assumption that space charge is zero everywhere. A self-consistent solution is sought where the net electric current of holes and electrons just inside the silicon must equal the current flowing in or out of the cell terminals, and the electric field integrated through the cell must equal the terminal voltage of the cell. The final results are printed out or plotted.

This program and others such as that of Fossum (1976) and the two-dimensional program written at Purdue University (Schwartz, Gray, and Lundstrom, 1982) can be applied to a wide variety of cell designs. As more is learned about carrier behavior in different semiconductors and at high concentrations, that new information can be incorporated in these programs. With a change of certain numerical values and internally stored data and equations, a program written for silicon can be converted to apply to gallium arsenide or other semiconductors. The flexibility of these programs, and the speed with which they enable one to try new cell designs, makes them invaluable tools in solar cell research and development.

ECONOMIC AND COMPLETE PV SYSTEM SIMULATIONS

SAMICS (Solar Array Manufacturing Industry Costing Standards) is a routine for calculating, with a computer program or as a manual procedure, the prices of manufactured PV modules. With it one can estimate prices that would result from use of particular fabrication techniques to make solar cells of given size using specified cell materials, electrodes, coatings, and encapsulants. SAMICS, accessed through the Jet Propulsion Laboratory (Pasadena, Calif.), has international applicability as only the standard data in it need be changed.

An example of a complete PV system program is that written at the Aerospace Corporation (1975), which yields estimates of the costs of solar electricity and its competitiveness with electricity from other sources to meet a time-varying demand. The modularity of this program is evident in Fig. A7.1. Historically correct electricity demand and direct solar insolation data may be entered via magnetic tape records. Each module or subroutine shown can be altered as one studies different concentrators, cells, or system configurations. Dashed lines in Fig. A7.1 represent flow of electricity from back-up generating plants into storage at night, and during or just before cloudy weather. Dotted lines show flow of power for array tracking and cell cooling.

Results of other simulations using different, possibly proprietary, PV system simulations have been reported in the *Proceedings of the IEEE Photovoltaic Specialists Conference*. Extensive results for PV and combined PV-thermal systems are tabulated in Volume 2 of OTA (1978). As a final example, the NASA-Lewis Research Center (Cleveland, Ohio), has developed a program that was used to design the Schuchuli village PV system (Chap. 7). The program calculates average PV cell output by month for a particular location. It also determines PV array size, tilt angle, and battery storage capacity to meet a given load profile. Inputs are average monthly values (EDS, 1968) of insolation, sky cover, precipitable water, and atmospheric turbidity, plus solar cell area, efficiency, and maximum-power voltage. The simulation includes provision for PV module degradation and random variations in insolation. Selection of the optimum array size and battery capacity combination can be modified by system considerations, such as physical limitations for the PV array or batteries, or by economic considerations.

280 COMPUTER SIMULATIONS OF PHOTOVOLTAIC CELLS AND SYSTEMS

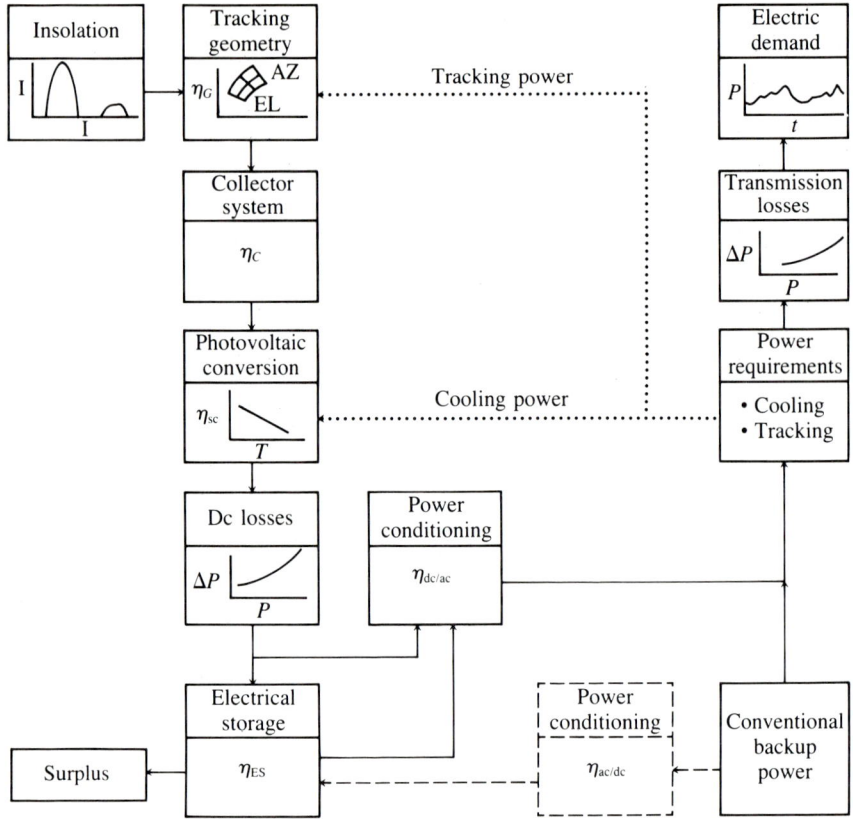

Figure A7.1 Photovoltaic system simulation model. In the plots, η = efficiency, AZ = azimuth, EL = elevation, I = direct solar insolation, P = power, ΔP = power dissipated resistively, T = temperature, and t = time.

DATA BASES AND SPECIAL ANALYSIS PROGRAMS

Data on the solar spectrum as obtained from measurements in space and corrected for absorption in the atmosphere have been published by ERDA (1977), as mentioned in Appendix 3. Measurements of intensity of sunlight have been made continuously for years at selected locations in the world. Questions about availability of insolation data for the United States should be directed to the DOE or SERI.

Standard electrical circuit analysis routines can be used to obtain I-V characteristics of lumped element equivalent circuits for solar cells. These routines range from circuit analysis modules run on small programmable calculators to large, fast computer programs such as SPICE (for information, contact the Publication Office, Electronics Research Laboratory, University of California, Berkeley).

Finally, in studies of concentrator cells having complex shapes, one may wish to model heat flow—conduction, convection, and a linear approximation to radiative transfer—with a heat transfer program, such as the two-dimensional finite-element Fortran program HEAT (for information, contact Prof. R. L. Taylor, Civil Engineering Dept., University of California, Berkeley).

REFERENCES

Aerospace Corp. (1975), Mission Analysis of Photovoltaic Solar Energy Systems, Final Report, ATR-76(7476-01)-1, Vol. 2, El Segundo, Calif., December.

Chappell, T. I. (1976), A pn Junction Silicon Sensor for High Intensity Solar Flux Mapping, Proc., 12th IEEE Photovoltaic Spec. Conf., 760–763, November.

Chappell, T. I. (1978), The V-Groove Multijunction Solar Cell, Proc., 13th IEEE Photovoltaic Spec. Conf., 791–796; updated version in *IEEE Trans. Elec. Dev.*, Vol. 26, 1091–1097, July 1979.

EDS (1968), *Climatic Atlas of the U.S.*, U.S. Environmental Data Services, Washington, D.C.

Fossum, J. G., (1976), Computer-Aided Numerical Analysis of Silicon Solar Cells, *Solid State Electr.*, Vol. 19, 269–277, April.

Gwyn, C. W., Scharfetter, D. L., and Wirth, J. L. (1967), The Analysis of Radiation Effects in Semiconductor Junction Devices, *IEEE Trans. Nucl. Sci.*, Vol. 14, No. 6, 153–169, December.

OTA (1978), Application of Solar Technology to Today's Energy Needs, Office of Technology Assessment, U.S. Government Printing Office, June and September.

Scharfetter, D. L., and Gummel, H. K. (1969), Large-Signal Analysis of a Silicon Read Diode Oscillator, *IEEE Trans. Elec. Dev.*, Vol. 16, 64–77, January.

Schwartz, R. J., Gray, J. L., and Lundstrom, M. S. (1982), "The intensity dependence of surface recombination in high concentration solar cells with charge induced passivation," Proc., 16th IEEE Photovoltaic Spec. Conf., September (to be published).

APPENDIX

8

SUPPLIERS OF SOLAR CELLS

Commercial suppliers of solar cells listed below are identified as follows:

[C-cell type] cells of type noted (Si, GaAs, CdS, conc.—for concentrator)
    [S] entire PV systems (cells and arrays, equipment for power conditioning and storage, system control and monitoring)
    [B] batteries for solar cell systems
    [PC] power-conditioning equipment
    [SP] cells for use in space environment
    [O] other solar cell products

AEG-Telefunken, Industriestrasse 29, D-2000 Wedel (Holst.), Germany [C-Si, poly.-Si][S][SP]

ANSALDO, Via N. Lorenzi 8, 16152 Genoa, Cornigliano, Italy [C-Si][S]

ARCO Solar, Inc., 20554 Plummer, Chatsworth, CA 91311 [C-Si][B][PC]

Applied Solar Energy Corp., P. O. Box 1212, City of Industry, CA 91749 [C-Si, conc.][SP]

British Aerospace Dynamics Group, Filton, Bristol, BS127QW, England [SP]

Central Electronics Ltd., Site 4, Industrial Area, Sahibabad, U.P. 201005, India

Edmund Scientific Co., 5975 Edscorp Building, Barrington, NJ 08007

Energy Conversion Devices, Inc., 1675 W. Maple Rd., Troy, MI 48084 [C-amorphous Si]

Hughes Aircraft Co., Box 92919—Airport Sta., Los Angeles, CA 90009 [SP]

Japan Solar Energy Co., Kyoto, Japan

Kyoto Ceramic Co., 52-11 Inouecho, Higashino, Yamashina-ku, Kyoto, 607 Japan [C-Si][S]

Lockheed Missiles & Space Co., P. O. Box 504, Sunnyvale, CA 94086 [SP]

Martin Marietta, M.S. C0470, P. O. Box 179, Denver, CO 80201 [C-Si, conc.][S][SP]

Matsushita Electric, Kadoma, Osaka, Japan

Mobile Tyco Solar Energy Corp., 16 Hickory Dr., Waltham, MA 02154 [C-Si][S]

Nippon Electric Co., Japan

Optical Coating Laboratories, Inc., 2789 Giffen Ave., Santa Rosa, CA 95401 [C-Si][O-AR coatings]

Phillips Gloeilampenfabrieken, Elcoma Dept., Eindhoven, Netherlands [C-Si]

Photon Power, Inc., 1067 Gateway West, El Paso, TX 79903

Photowatt Intl., Inc., 2414 W. 14th St., Tempe, AZ 85281 [C-Si][S]

Poly Solar Inc., 2701 National Dr., Garland, TX 75041

Radio Shack, retail outlets throughout United States [C-Si]

RTC, Route de la Delivrande, 14000 Caen-Cedex, France

SEMIX, Inc., 15809 Gaither Rd., Gaithersburg, MD, 20760 [O-semi-crystalline silicon sheet for cells]

SES, Inc., Tralee Industrial Park, Newark, DE 18711 [C-CdS]

Sharp, 2613.1 Ichinomoto, Tenri-Shi, Nara, Japan [C][SP]

Solar Electric International, 4837 Del Ray Ave., Washington, D.C. 20014 [O-PV powered irrigation pumps]

Solar Generator, Singapore Pte. Ltd., 151 Loroug Chuan, Singapore, 1955 [C-Si]

Solar Power Corp. 20 Cabot Rd., Woburn, MA 01801 [C-Si][S]

Solar Usage Now, Inc., 450 E. Tiffin St., Basom, OH 44809 [C-Si][S]

Solarex Corp., 1335 Pickard Dr., Rockville, MD 20850 [C-Si][S][PC][B]

Solavolt International, 3646 E. Atlantic Ave., Phoenix, AZ 85062 [C-Si][S]

Solec International, Inc., 12533 Chadron Ave., Hawthorne, CA 90250 [C-Si][S]

Solectro-Thermo, Inc., 1934 Lakeview Ave., Dracut, MA 01836 [O-total energy systems (electrical and thermal output)]

Solenergy Corp., 171 Merrimas St., Woburn, MA 01801 [C-Si][S]

Spectrolab, Inc., 12500 Gladstone Ave., Sylmar, CA 91342 [C-Si, GaAs][S][SP]

Spire Corp., Patriots Park, Bedford, MA 01730 [C-Si, GaAs]

Tideland Energy Pty. Ltd., P. O. Box 519, Brookvale, N.S.W. 2100, Australia

Tideland Signal Corp., P. O. Box 52430, Houston, TX 77052 [S-solar-powered navigation aids]

Varian Associates, 611 Hansen Way, Palo Alto, CA 94306 [C-GaAs, conc.]

APPENDIX

9
OVERVIEW OF SOME OPERATING SYSTEMS

The following tables summarize the operating characteristics of a number of remote stand-alone, residential, and intermediate load center PV systems (from J. L. Rease, "Photovoltaic Systems Overview", 14th IEEE Photovoltaic Spec. Conf., pp. 1139–1145, 1980).

Table A9.1 Remote stand-alone applications

User/application/ quantity/total watts/instal. date	Service	Operational period	Status
IHS/refrigerator/1/ 330/July 1976	Used by Papago Indian village for medical & food supplies at Sil Nakya, AZ.	Continuous	A 12-ft^3 Magna-Kold refrigerator is being readied to replace the 4-ft^3 recreational vehicle refrigerator which has excessive run time in hot weather.
USFS/forest lookouts/2 588/October 1976	Antelope Peak, CA, and Pilot Peak, CA. Power for water, lights, radio and refrigerator.	May to October	Both lookouts closed for winter in late October. No problems. Dow Butte Lookout (north of Antelope Peak) equipped with PV system in October.
NOAA/weather station/5/ 481/April-August 1977	Halfway Rock, ME; Loggerhead Key, FL; Clines Corners, NM; Pt. Retreat, AK; South Pt., HI; RAMOS weather station power supply.	Continuous	Halfway Rock, ME, system became operational 9/25. One of the 3 strings in the PV array is noncontributing.
DOT/highway sign/1/116/April 1977	Power for receiver, transmitter, motor, and lights on changeable message, dust warning, sign near Tucson, AZ, on I-10.	Continuous	In operation.
USDA/insect survey traps/4/ 372/May 1977	Near Texas A & M in College Station, TX. Power for two black light and two electric grid insect survey traps.	April through September	Co-op experiment period completed. Disposition under review.

Lone Pine, CA/H$_2$O cooler/1/ 446/September 1977	Demonstration PV-powered water cooler for Owens Valley Interagency Committee Visitor Center, Lone Pine, CA.	Continuous	In operation.
NPS/refrigerator/1/ 220/July 1978	Food storage for back country ranger station in Isle Royale National Park, MI, wilderness area.	Summer 1976 Summer 1978 Summer 1979 Summer 1980	System removed for winter.
IHS-PTC/village power/1/ 3500/December 1978	Provide power to 95 people in Papago Indian village at Schuchuli, AZ, for water pump, 15 refrigerators, 47 lights, washing machine, and sewing machine.	Continuous	In operation.
NJ-DEP/air quality monitor/ 1/360/Nov. 1979	Demonstrate PV power for operation of high volume air sampler at Liberty State Park, NJ.	Continuous	In operation.
USGS/seismic sensors/1/40/Jan. 1980	Provide power to operate seismic instruments for Hawaii Volcanic Observation in Hawaii.	Continuous	In operation.

Table A9.2 Residential systems

Project	Operational	Array	Array code*	Inverter
Univ. of Texas/Arlington	Nov '78	6.2 kW$_{pk}$ Sensor Tech Retrofit	G	Gemini 3 kVA
John F. Long Fiesta	June '80	4.5 kW$_{pk}$ ARCO Solar	D	Gemini 8 kVA
Carlisle, MA ISEE	Feb '81	7.3 kW$_{pk}$ Solarex	SO	Gemini 8 kVA
Molokai Wiepke House	Apr '81	4 kW$_{pk}$ ARCO Solar	SO	Gemini 4 kVA
Pearl City	Apr '81	4 kW$_{pk}$ ARCO Solar	SO	Gemini 4 kVA
Kalihi	Apr '81	2 kW$_{pk}$ ARCO Solar	SO	Gemini 2 kVA
Florida Solar Energy Center	Nov '80	5 kW$_{pk}$ ARCO Solar	SO	2-Gemini 4 kVA
MIT/LL NE Prototype	Dec '80	6.9 kW$_{pk}$ Solarex	SO	Gemini 8 kVA
GE NE Prototype	Mar '81	6.1 kW$_{pk}$ GE Shingle	D	Abacus 6 kVA
Solarex NE Prototype	Feb '80	5.3 kW$_{pk}$ Solarex	SO	Abacus 6 kVA
TriSolar Corp. SW Prototype	Apr '81	4.8 kW$_{pk}$ ASEC	I	Gemini 8 kVA
Westinghouse NE Prototype	Feb '81	5.4 kW$_{pk}$ ARCO Solar	I	Abacus 6 kVA
TriSolar Corp SW Prototype	Apr '81	4.6 kW$_{pk}$ ASEC	I	Gemini 8 kVA
Westinghouse SW Prototype	Apr '81	5.4 kW$_{pk}$ ARCO Solar	I	Abacus 6 kVA
GE SW Prototype	June '81	5.6 kW$_{pk}$ GE Shingle	D	Abacus 6 kVA
ASI SW Prototype	TBD	5.9 kW$_{pk}$ ARCO Solar	D	Gemini 8 kVA
BDM SW Prototype	Apr '81	4.3 kW$_{pk}$ Motorola	SO	Abacus 6 kVA
ARTU SW Prototype	TBD	4.9 kW$_{pk}$ ARCO Solar	SO	TBD
Solarex SW Prototype	Apr '81	4.8 kW$_{pk}$ Solarex	SO	Abacus 6 kVA
TEA SW Prototype	Apr '81	3.7 kW$_{pk}$ Motorola	R	Abacus 6 kVA

* Array code: G—ground mount, D—direct mount, SO—stand-off mount, I—integral mount, R—rack mount.

Table A9.3 Intermediate load center systems

Project	Operational	Array	Inverter	Buyback
Mead, Nebraska	Jul '77	13.5 kW$_{pk}$ Solarex 1.46 kW$_{pk}$ Sensortech	NOVA 3 @ 10 kVa	None
Bryan, Ohio	Aug '79	15 kW$_{pk}$ Solarex	DC	None
Mt. Laguna, California	Aug '79	49.5 kW$_{pk}$ Solar Power 15.1 kW$_{pk}$ Solarex	DECC 75 kVA	None
Natural Bridges National Monument, Utah	Jan '80	47.3 kW$_{pk}$ Motorola 17.9 kW$_{pk}$ Spectrolab 31.3 kW$_{pk}$ ARCO	Cyberex 40 kVA	None
Newman Station, El Paso, Texas	Dec '80	17.5 kW$_{pk}$ Solar Power	DC	None
Mississippi Country Community College, Blytheville, Arkansas	May '81	240 kW$_{pk}$ Solar Kinetics Parabolic Trough-Solarex Cells	DECC 300 kVA	100%
Northwest Mississippi Junior College, Senatobia, Mississippi	Aug '81	50 kW$_{pk}$ UTL GaAs 50 kW$_{pk}$ Team 6X 50 kW$_{pk}$ SKI-Solarex 1 kW$_{pk}$ Photon Power CdS	Windworks 300 kVA	None
Lovington, New Mexico	Apr '81	100 kW$_{pk}$ Solar Power	DECC 2 @ 60 kVA	None

(*Continued*)

Table A9.3 (continued)

Project	Operational	Array	Inverter	Buyback
Beverly HS, Beverly, Massachusetts	May '81	100 kW$_{pk}$ Solar Power	DECC 2 @ 60 kVA	100%
Oklahoma Museum of Science & Art, Oklahoma City, Oklahoma	Aug '81	135 kW$_{pk}$ Solarex (reflector)	Windworks 150 kVA	100%
BDM Office Building, Albuquerque, New Mexico	June '81	47 kW$_{pk}$ SKI Parabolic Trough—ASEC Cells	Westinghouse 62.5 kVA	None
Wilcox Hospital, Lihue, Kauai, Hawaii	Aug '81	35 kW$_{pk}$ Acurex Parabolic Trough—ASEC Cells	Westinghouse 62.5 kVA	None
Dallas-Fort Worth Airport	Nov '81	27 kW$_{pk}$ E-Systems Linear Fresnel—ASEC Cells	Jim Ross & Assoc. 27 kVA	None
Sea World, Orlando, Florida	Dec '81	110 kW$_{pk}$ GE Parabolic Trough	Windworks 150 kVA	None
Sky Harbor Airport, Phoenix, Arizona	Dec '81	225 kW$_{pk}$ Motorola PF Fresnel	Power Systems & Controls 250 kVA	None
San Bernardino West Side Community Dev. Corp. Concrete Plant, San Bernardino, California	Oct '81	35 kW$_{pk}$ Solarex	DECC 75 kVA	100%

INDEX

Boldface type is used to indicate the page on which a term is defined.

a-Si:F:H, 208–210
a-Si:H, 208–210
AM1 sunlight, 73
Absorption coefficient, **45**, 204–207, 215
Acceptance angle, **98**
Acceptor, **41**
Air mass number, **21**, 264
AlAs, 203
AlGaAs, 268–269
AlSb, 203
Ambient, effect on cells, 72
Amorphous materials, **73–75**
Amorphous silicon (*see* Cell types, amorphous silicon)
Annealing, 72, 76, 193–195
Anodization, 196
Antireflection coating, 70, 81–83, 196, 199, 212, 213–219, 264
Applications:
  airplane, 137
  alarms, 137
  automobile, 174–175
  cattle fence, 137
  central power station, 136
  communications, 136, 148–149
  consumer electronics, 137
  instrumentation, 286–287
  irrigation, 137, 146–147
  lighting, 136, 143
  navigational aids, 133
  radio, 136, 139, 151, 286–287
  railroad, 136
  refrigeration, 137, 143, 286–287
  remote, 147–150, 286–287
  repeater station, 136, 149
  residential, 139, 288
  television, 136, 149
  village power system, 136, 139–144, 286–287
  water pumping, 136, 143, 146–147
  water purification, 136, 148
Array (*see* Concentrator)
Array fabrication, 83
Auger recombination, 216
Automated processing, 90

Back-surface field (BSF), **65**, 193, 264
Balance-of-system (BOS), 72, 138, 140, 158, 264
Bandgap, **40**, 206–207
  direct/indirect, **46**
Bandgap narrowing, 216
Barrier layer, 81

Battery, 115–116, 124–125
  lead-acid, 124, 143, 145
  redox, 125
Blocking diodes, 143, 145, 217
Bridgman process, 182–184
Butyl rubber, 197
Buy-back, utility, 141, 165–171, 173

Carrier:
  majority, **41**
  minority, **41**
Cathodic protection, 133, 137
CdS (*see* Cell types, cadmium sulfide)
CdSe, 203, 214, 220, 268–269
CdSe, 268–269
CdTe, 203, 204, 214, 220, 268–269
CdZnS/Cu$_2$S, 214, 217, 220, 268–269
Cell, concentrator (*see* Cell types, concentrator)
Cells, commercial availability, 129, 282–284
Cell types:
  amorphous silicon, 9, 151–152, 192, 203, 204, 208–210, 214, 220, 268–269
  cadmium sulfide, 85–86, 131, 192, 203, 204, 210–211, 220, 268–269, 289
  cascade, 242
  concentrator, 108–109, 227–235, 238–247, 289–290
  dendritic web, 186–189
  discontinuous junction, 226
  edge-defined film-fed growth (EFG), 186–189, 197, 265, 268–269
  flat-plate, 129–131
  front-surface-field, 226
  gallium arsenide, 9, 129, 131, 203, 204, 220, 268–269, 289
  graded-bandgap, 233
  heterojunction, 224
  high-low junction, 230
  induced junction, 225
  interdigitated-back-contact, 230
  liquid-junction, 227
  MIS, 10, 209, 225, 265, 268–269
  MOS, 10, 265
  multiple pass, 227
  multiple-junction, 227–230
  n-on-p, 79
  organic, 9, 203, 213, 214, 220, 268–269
  photoelectrochemical, 8, 9, 192, 247–251, 266, 268–269
  p-i-n, 209–210, 268–269
  photoelectrolytic, **247–251**
  photogalvanic, **248**

291

Cell types (*continued*):
  pn junction, 10, 79
  polycrystalline, 9, 131, 134–135, 184–185
  ribbon, 9, 180
  Schottky-barrier, 10, 196, 205, 212–213, 224
  semicrystalline, 9, 268–269
  silicon, 70, 129–131, 203, 204, 214, 220, 268–269, 286–290
  spectrum splitting, 238–242
  tandem-junction, 226
  thermophotovoltaic, 9, **244–247**
  thin film, 8, 213–220, 268–269
  V-groove multijunction, 230
Chemical vapor deposition (CVD), 182, 191, 212, 264, 268–269
Chopper, 116
CLEFT process (*see* Epitaxy)
Collection efficiency, **56**
Collector:
  tilt, **35**
  tracking, 19–20
Concentration ratio, **96**
Concentrator, 8, 96–106
  compound parabolic, **104**
  fresnel lens, **103**
  holographic, **105**
  imaging/nonimaging, **97**, 104
  luminescent, **105**
  one/two-dimensional, **97**
  parabolic dish, **105**
  parabolic trough, 101
  truncated compound parabolic, **104**
Concentrator arrays, 133, 139, 141, 149–150
Concentrator cells, 8, 9, 108–109, 129–130, 132–133, 227–235, 238–247, 268–269, 289–290
  See also Cell types, concentrator
Concentrator systems, economics of, 106–108
Contacts, 205, 213, 215
Cooling, 110–111
Copper sulfide, 85
Cost:
  balance-of-system, 160–161
  of concentrator cells, 11, 13, 129
  of electricity, 158
  incremental, 169
  levelized, 159, 175
  life-cycle, 165–168
  marginal, 172
  of solar cells, current, 11, 13, 129, 132, 140, 146, 158, 160–161
  of solar cells, goals, 11, 13, 79, 91, 161–165, 197–199, 218–220
Cover glass, 83
Crystallinity:
  of cell material, 9, 73
  semicrystalline cells, 8, 9, 73
Cu_2O, 203, 204, 220
Cu_2S, 203, 204, 219, 268–269
Cu_2Te, 203, 220, 268–269
$CuInS/CuInS_2$, 203, 204, 214, 268–269

$CuInSe_2$, 203, 204, 210–211, 220, 268–269
Current, short-circuit, **54**
Czochralski (CZ) process, 77, 182–184, 264

Damage, 76, 193
Defects, 74, 186, 208
Degeneracy, 216
Dendritic web, 9, 168–169
Depletion region, **50**, 52
Developing nations, 86, 173–174, 265
Diffusion, 79, 201, 205, 217
Diffusion coefficient, **48**
Diffusion length, **48**, 205, 208–209, 215
Diode, **53**
Dipping, 85
Donor, **41**
Dopant, **40**
Doping, 71, 79–80, 193
Doubling time, 15
Drive-in, 80

Economic feasibility, 133, 155–158
Efficiency:
  of cells, 129, 157–158, 268–269, 271–275
  collection, **56**
  conversion, **59**, 60–67, 129, 157–158, 268–270
  effect of bandgap, 60, 203
  effect of intensity, 63, 108–109
  effect of temperature, 61–62, 110
Electric utility:
  buy-back by, 141, 165–171, 173
  connection to, 118–121, 125, 169, 171–173
Electrodeposition, 89, 192, 196, 268–269
Electrodes, 81–82
Electrolysis, 8
Electron affinity, 215, 217
Electronresist, 71
Encapsulant, 90, 196–197, 217, 219
Energy:
  demand, 11, 12, 163–164, 170, 279–280
  efficiency of utilization, 11, 12, 15
Energy-band diagram, **42**
Energy content, of materials, 87–88
Energy gap, **40,** 203–207, 215, 275
Energy payback ratio, 72
Energy payback time, **15**, 72, 79–80, 199
Energy storage, 8, 122–126
Energy use, 12, 153, 157, 260
Environmental effects, 73, 86–87, 144
Epitaxy, 265
  CLEFT process, 190, 268–269
  liquid-phase (LPE), 182, 190, 265
  molecular-beam (MBE), 190–191, 265
  solid-phase (SPE), 182, 266
  vapor-phase (VPE), 182, 190, 267
Equation of time, **99**
Equinox, 18
Etching, 83
Evaporation, 81, 191–192, 208, 268–269
Experiments, **271–275**

INDEX

f-number, **98**
Failure (*see* Reliability)
Field-assisted bonding, 197
Fill factor (FF), **59**, 265, 268–269
Fixed-charge rate, 159–160
Flat-plate cells, 9
Float-zone (FZ) process, 77, 182–184
Fossil fuel, 6, 11, 13, 86–87
Fresnel lens, **103**
Fusion, 17

GaAlAs, 190
GaP, 203
GaAs or gallium arsenide (*see* Cell types, gallium arsenide)
Ge, 203
Gettering, 193
Glass:
    float, as substrate, 86, 89, 180, 192
    low iron content, 84
Grain boundaries, 74, 86
Graphoepitaxy, 191
Growth:
    of crystals, 77, 182–184
    exponential, 15
    logistic, 14

HEAT computer simulation, 281
High-level injection, **53**
Hole, **40**
Hydrazine, 83

InP, 203, 210–212, 214, 220, 268–269
InSe, 203, 214
Index of refraction, 74–75, 82–83
Indium-tin oxide (ITO), 196, 212, 213, 265, 268–269
Injection, carrier, **50**
Insolation (*see* Solar radiation)
Integrated circuit (IC), 70–72, 265
Intensity, of sunlight, 6, 11, 157, 260, 280
Intermediate load center, 158–165, 289–290
Intrinsic carrier concentration, **41**
Inverter, 118–121, 141, 288–289
    and power grid, 118–121
    stand-alone, 118
Ion implantation, 72, 76, 193–195
I-V characteristic, 271–275

Junction, *pn*, **49**

Kerf loss, 77

Langley, 260
Laser processing, 193–196
Lattice mismatch, 216
Leaching, 196
Lens, of camera, 82, 98
Less-developed countries (LDC), 86, 265
Levelized cost, 159, 175
Lifetime:
    minority carrier, **46**, 74, 76, 193, 195, 205, 275
    recombination (*see* Lifetime, minority carrier)
Lightning, 140
Losses, 216

Maintenance, of PV system, 153, 158–159
Market fraction, 14
Materials, availability of, 72, 90
Maximum-power point, **59**, 117
Mead, NB agricultural system, 138, 144–146
Merocyanine, 203, 214, 220, 268–269
MG-Si, 77–78
Microirrigation, 146–147
Microwave lifetime measurement, 76
Migration 86
$MoSe_2$, 268–269
Mobility, **42**, 205
Modularity, 11
Modules, 130–132

Navigational aids, 133

Ohmic contact, 81
Oil, 6
Open-circuit voltage, **58**, 271–275
Optical absorption coefficient, **45**, 204–207, 215
Organic solar cells (*see* Cell types, organic)
Ownership, of PV system, 153
Oxidation, 71

PERSPEC computer simulation, 277–279
Phase diagram, 183
Photoconductor, **57**
Photodiode, **57**
Photoelectrochemical (PEC) cells, 8, 9, 192, 247–251, 266, 268–269
Photoelectrolytic system, 247–251
Photolithography, 71, 77, 191
Photon energy, **42**
Photoresist, 71–72, 266
Photosynthesis, 81, 213
Photovoltaic cloth, 186
Photovoltaic effect, 6, 8, 39
Photovoltaic shingle, 150, 288
Plasma deposition, 192
Plasma etching, 71
Polycrystalline materials, 75, 181, 191–192, 196
Polymethylmethacrylate, 84
Polyvinylbutyral, 196–197
Power conditioning, **114**, 121–122
Power, maximum, 271–274
Predeposition, 79
Printing, of contacts, 219
Production rate, of cells, 10, 13, 88–89, 90, 201
Properties, of semiconductors, 74, 204–207, 221
Pumped hydro storage, 125–126
PURPA, 196
Pyrheliometer, **24**

Pyranometer, 23, **24**
   shading-ring, 23, **24**

Quad, 11, 259
Quartzite, 77

Rapid-quenching process, 185, 187–190, 268–269
Recombination, **46**, 76, 86, 186, 216
   Auger, **48**
Recombination center, **48**
Recrystallization, 195–196
Rectifier, **53**
Reliability, 138, 140
Residential lighting, 136
Reverse saturation current, **51**
Ribbon-against-drop growth process, 192
Ribbon cells (*see* Cell types, ribbon)
Ribbon, self-supporting, growth of, 185–190
Ribbon-to-ribbon regrowth (RTR), 188–190, 266
RTV, 84

SAMICS computer simulation, 197, 218–219, 279
Satellite power system, 251–254
Schottky junction, 53
Schuchuli, AZ project, 142, 287
Semiconductor, **40**
   intrinsic, **41**
Semiconductor grade, 77–78
Semiconductor-insulator-semiconductor (SIS), 212
Series-parallel connection, 84, 144–145
Series resistance, **65–66**, 81, 109, 216–217, 227–232
Shingle, photovoltaic, 150, 288
Short-circuit current, **54**, 271–275
Silicon (*see* Cell types, silicon):
   metallurgical grade (MG-Si), 77–78, 184–185
   semiconductor grade (SeG-Si), 77–78
Silicon carbide, 209
Silicone, 84
Silicon-on-ceramic, 187–188, 192, 266, 268–269
Silk-screening, 81, 196, 199
Simulation, computer, 144, 164–168, 197, 218–219, **276–281**
$(SN)_x$ polymer-semiconductor cell, 212–213
Solar altitude, 21
Solar breeder, 195, 199–200
Solar concentrator (*see* Concentrator)
Solar constant, **18**
Solar flux (*see* Solar radiation)
Solar noon, **19**
Solar power satellite (*see* Satellite power system)
Solar radiation, 20–34
   AM1, 22
   geographical distribution, 28–34
   map, U.S., 28, 30, 31

   map, world, 32, 33
   measurement, 23–26
   seasonal variation, 29, 34
Solar simulator, 85
Solar spectrum, 23

   AM1.5, 262–263
Solar tracking, 19, 28, **99**
Soleras project, 139, 141
Solidification, 181
Solstice, 18
Space power system (*see* Satellite power system)
Space-charge region, **50**
Specification sheet, 134–135
Spectral response, **56**
Spectrum, solar, 262–263
Spectrum splitting, 238–242
SPICE computer simulation, 281
Spraying, 196, 199, 212, 219
Sputter deposition, 89, 192, 208, 212
Stability, of cells, 210
Stepanov growth process, 186–187
Storage, battery, 143, 144, 172
Sun, -earth motion, 17–20
Suppliers of cells and PV components, 282–284
SUPREM computer simulation, 277
Surface recombination velocity, **51**, 65, 226

Tandem cell, 242–244
Tax credit, 156
Temperature effects, 217, 271, 275
Texturing, 83–84
Thermal activation, 181–182, 201
Thermal conductivity, 205
Thermal energy, 110–111
Thermal expansion, 205
Thermophotovoltaic (TPV), 9, **244–247**, 266
Time-of-day pricing, 175
Tin oxide, 212
Total energy systems, 139, 149–150, 153
Tracking (*see also* Solar tracking):
   maximum-power-point, **114–115**, 116–118
Transients, power line, 172
Trap, **48**
Trichlorosilane, 77
Tunneling, 217

Uninterruptible power supply (UPS), 146
Units, 259–261
Utilization factor, 159–160

Voltage regulation, 143

Wafering, 77–78, 180
Winston collector, **104**
WSe_2, 268–269

Zn_3P_2, 203, 204, 220, 268–269
ZnP_2, 214
ZnTe, 203